GENETIC INFLUENCES ON HUMAN FERTILITY AND SEXUALITY

Theoretical and Empirical Contributions from the Biological and Behavioral Sciences

GENETIC INFLUENCES ON HUMAN FERTILITY AND SEXUALITY

Theoretical and Empirical Contributions from the Biological and Behavioral Sciences

edited by

Joseph Lee Rodgers
University of Oklahoma

David C. Rowe
University of Arizona

Warren B. Miller
Transnational Family Research Institute

KLUWER ACADEMIC PUBLISHERS
Boston / Dordrecht / London

Distributors for North, Central and South America:
Kluwer Academic Publishers
101 Philip Drive
Assinippi Park
Norwell, Massachusetts 02061 USA
Telephone (781) 871-6600
Fax (781) 871-6528
E-Mail <kluwer@wkap.com>

Distributors for all other countries:
Kluwer Academic Publishers Group
Distribution Centre
Post Office Box 322
3300 AH Dordrecht, THE NETHERLANDS
Telephone 31 78 6392 392
Fax 31 78 6546 474
E-Mail <services@wkap.nl>

 Electronic Services <http://www.wkap.nl>

QP
251
.G353
2000

Library of Congress Cataloging-in-Publication Data

Genetic influences on human fertility and sexuality: theoretical and empirical
contributions from the biological and behavioral sciences / edited by Joseph Lee
Rodgers, David C. Rowe, Warren B. Miller.
    p ; cm.
  Includes bibliographical references and index.
  ISBN 0-7923-7860-1 (alk. paper)
    1. Fertility. 2. Human reproduction. 3. Human genetics. I. Rodgers, Joseph Lee. II.
Rowe, David C. III. Miller, Warren B., 1935-
  [DNLM: 1. Fertility-- genetics. 2. Genetics, Medical. 3. Reproduction--genetics. 4.
  Sexuality. WP 565 G328 2000]
  QP251 .G353 2000
    621.6--dc21

                                                                                    00-037111

Printed on acid-free paper.   Printed in the United States of America

CONTENTS

CONTRIBUTORS

Christine Bachrach, Demographic and Behavioral Sciences Branch, Center for Population Research, NICHD, 610 Executive Blvd, Room 8B13, Bethesda MD 20892, bachracc@exchange.nih.gov

J. Michael Bailey, Department of Psychology, Northwestern University, 2029 Sheridan Road, Evanston, IL 60208-2710, jm-bailey@nwu.edu

Jay Belsky, Department of Psychology, Birkbeck College, University of London, Malet Street, London WC1E 7HX, United Kingdom, j.belsky@psychology.bbk.ac.uk

Hezekiah Blake, 4136 Alabama St., San Diego, Ca, 92104, hblake@ucsd.edu

Mary H. Burleson, Department of Social and Behavioral Sciences (MC 3051), Arizona State University West , P.O. Box 37100, Phoenix, AZ 85069, Mary.Burleson@asu.edu

Kaare Christensen, Epidemiology, Institute of Public Health, University of Southern Denmark: Main campus Odense University, Winsloewparken 17, DK-5000 Odense C, Denmark, k-christensen@win-chs.ou.dk

David E. Comings, Department of Medical Genetics, City of Hope Medical Center, Duarte, CA 91010, dcomings@mindspring.com

Khytam Dawood, Department of Psychology, Northwestern University, 2029 Sheridan Road, Evanston, IL 60208-2710, khytam@nwu.edu

Debby Doughty, Department of Psychology, University of Oklahoma, Norman OK 73019, ddoughty@ou.edu

A. J. Figueredo, Department of Psychology, University of Arizona, Tucson, AZ 85721, ajf@u.arizona.edu

Irving I. Gottesman, Department of Psychology, University of Virginia, Gilmer Hall, Charlottesville VA 22903, iig@virginia.edu

Dean H. Hamer, Laboratory of Biochemistry, National Cancer Institute, National Institues of Health, Bethesda, MD 20892, hamerd@dc37a.nci.nih.gov

Kimberly A. Hughes, Department of Life Sciences (MC 2352), Arizona State University West, P.O. Box 27100, Phoenix, AZ 85069 Kimberly.Hughes@asu.edu

Patrick Johnson, Psychiatry Service (116A1), Pettis Memorial Veterans Administration Medical Center, Loma Linda, Ca, 92357, jpjohn@earthlink.net

Hans-Peter Kohler, Max Planck Institute for Demographic Research Doberaner Str. 114, 18057 Rostock, Germany, kohler@demogr.mpg.de

Bruce Kovacs, Department of OB/GYN, room SK-40, University of Southern California, LAC & USC women's & Children's Hospital, 1240 N. Mission Ave, Los Angeles, CA 90033, bwkovacs@uniphys.com

Irving I. Gottesman, Department of Psychology, University of Virginia, Gilmer Hall, Charlottesville VA 22903, iig@virginia.edu

James MacMurray, Vice President, Genomics Division, Medical Genetics, Inc., 1115 No. Yale Ave, Claremont, CA 91711, macmurr1@gte.net

Matt McGue, Department of Psychology, University of Minnesota, Elliott Hall, rm N460, 75 East River Road, Minneapolis, MN 55455-9991, mmcgue@tfs.psych.umn.edu

Linda Mealey, Psychology Department, College of St. Benedict, St. Joseph, MN 56374, lmealey@csbsju.edu

Warren B. Miller, Transnational Family Research Institute, 355 West Olive Avenue, Suite 207, Sunnyvale, CA 94086-7660, tfri@t-three.com

Donn Muhleman, Department of Medical Genetics, City of Hope Medical Center, Duarte, CA 91010, dmuhleman@coh.org.

David J. Pasta, P.O. Box 881, Palo Alto, CA 94302, dpasta@lewin.com

Joseph Lee Rodgers, Department of Psychology, University of Oklahoma, Norman OK 73019, jrodgers@ou.edu

David C. Rowe, Division of Family Studies, 210 FCR Building, University of Arizona, Tucson, AZ 85721 dcr091@ag.arizona.edu

Susan Trumbetta, Department of Psychology, Box 219 Vassar College, 124 Raymond Ave., Poughkeepsie, NY 12604-0219 trumbetta@vassar.edu

J. Richard Udry, Carolina Population Center, University Square, CB#8120, UNC-CH, Chapel Hill, NC 27599, udry@unc.edu

Jane H. Underwood, Department of Anthropology, University of Arizona, 2228 East 4th Street, Tucson, AZ 85719 kammagar@prodigy.net

PREFACE

We hope that the publication of this volume signals the beginning of a new and active research agenda, an agenda focused on the link between genes and human reproduction. Such research will necessarily be characterized by two important "I words": *Integrative* and *Interdisciplinary*. There have certainly been individual scientists in the past who have considered whether genetic influences can and do influence fertility and other fitness traits. But there has been little *integrative, interdisciplinary* effort to address the fascinating questions that arise from this consideration. Within our current research culture scholars routinely ask, "To what extent do genes influence _____?" The _____ can be filled in with dozens of interesting and socially-important topics, including disease, intelligence, personality, divorce, television viewing, and general health and happiness, just to name a few. One relatively unexplored set of behaviors, however, are the theoretical starting point for human existence: Those related to reproductive success. Surprisingly, little coherent attention has been given to whether and how genes influence sexual behavior, fertility, and reproduction. Likewise, there has been little coherent effort devoted to studying the combination of influences in gene/environment interactions. This volume is an effort to both provide and stimulate the development of that coherence.

In December, 1997, at the Arizona Inn in Tucson, in a charming little conference room furnished with a roaring fire and a heavily-decorated Christmas tree, two dozen scholars from all over the world convened a conference entitled "Genetic Influences on Fertility and Related Processes." The conference was co-sponsored by the Office of Behavioral and Social Science Research of the National Institutes of Health and by the Society for the Study of Social Biology. This edited volume emerged from that conference. A follow-up conference on the same topic is scheduled to occur in Rostock, Germany in summer, 2000, at the Max Planck Institute for Demographic Research; this second conference is also sponsored by the OBSSR branch of NIH, along with the Max Planck Institute.

Many individuals contributed time, energy, and insight to the production of this book. All of the papers were carefully peer-reviewed; the review team included Kaare Christensen, Bruce Ellis, Warren Miller, David Pasta, Joe Rodgers, and David Rowe. At the University of Oklahoma, Debby

Doughty provided a great deal of editing and formatting, and Ronnie Coleman assisted with production of the index. At Kluwer, Charles Schmieg, Mary Panarelli, and Joanne Tracy provided excellent support for the project. Financial support for the development of this book came from NIH Grant #R01-HD2-1973, which supported all three editors and Debby Doughty. Individuals at NIH who provided both financial and scholarly support included Christine Bachrach, Susan Newcomer, and Virginia Cain.

The editors would also like to acknowledge the support of their families, whose contributions are of a different type, but equally indispensable. To Jacci, Rachel, and Naomi Rodgers; to Carol and Scott Rowe; and to Casey Douglas and Kyle Miller we express our love, affection, and appreciation.

Finally, we join with past, present, and future contributors to encourage additional research in the highly interdisciplinary research arena studying both genetic and environmental influences on fertility and other fitness traits in humans. This effort promises the excitement of asking new, difficult and unusual questions; the stimulation of an intellectual community that unites in the pursuit of common goals; and the satisfaction achieved by taking small steps forward along the intellectual path that leads to better understanding of biological and social processes, and, especially, their interplay.

Joseph Lee Rodgers
David C. Rowe
Warren B. Miller

GENETIC INFLUENCES ON HUMAN FERTILITY AND SEXUALITY: INTRODUCTORY COMMENTS FROM A BIOLOGICAL SCIENTIST

Dean H. Hamer

Sexuality and fertility are at the nexus of evolution. The one thing that we can be absolutely sure about all of our ancestors is that they mated and had children. It is surprising, then, that the genetic analysis of these traits in humans has been so long relegated to the back burner of the research stove. For example, a computer scan of the behavioral genetics literature finds almost 100-fold more articles on personality traits than on sexual behavior.

This volume on "Genetic Influences on Human Fertility and Sexuality" begins to ameliorate this sad state of neglect. Based on a conference held in Tucson, Arizona in December, 1997, it contains articles and commentaries on a variety of topics that are central to human reproduction. Included are physical phenomena such as menarche and pubertal development, psychological characteristics such as sexual orientation, complex psycho-behavioral traits such as childbearing motivation and fertility behavior, twinning, and population genetics. As befits the broad range of topics, the authors come from many different branches of science including genetics, molecular biology, psychology, psychiatry, economics, and anthropology. It appears to represent the first time that so many different disciplines have been brought together to consider this important set of topics.

Why has it taken so long for such a volume to appear? Part of the reason is that many scientists in the past interpreted Fisher's Fundamental

Theorem of Natural selection to imply that there could be no genetic variance in fitness traits. It is now recognized, however, that Fisher was referring only to selection, and that mutation, genetic drift and pleiotropy can generate ample variation in the relevant loci. The proper question is not whether the genes related to fertility and sexual behavior *should* be variable, but whether they *are* variable. Fortunately, scientists can now empirically approach this question at both the population level, by studying twins and adoptees, and at the molecular level, by looking at DNA sequences. Both strategies are amply represented in this volume.

The second reason for the neglect of research on the genetics of human sexuality and fertility is political rather than scientific. Many academics, administrators and funders still consider sexuality too private, too controversial or just too plain embarrassing to be a topic for "respectable" scientists to work on. At the same time, research on childbearing and fertility behavior is tainted by the faint whiff of eugenics with which such work was all too often associated in the past.

Within the next year or two, we will know the complete sequence of the human genome and all of the 100,000 or so genes that make up our genetic blueprint. It can confidently be predicted that many of these genes — perhaps more than half — will be involved in the development and functioning of the brain and therefore in behavior, including sexuality, contraception, and childbearing. It would be sad not to learn what these genes are doing for reasons as silly as embarrassment or historical mistakes. Perhaps the greatest contribution of this innovative volume is that it demonstrates that genetic influences on human fertility and sexuality are an important and approachable topic for rigorous and thoughtful scientific research.

GENETIC INFLUENCES ON HUMAN FERTILITY AND SEXUALITY: INTRODUCTORY COMMENTS FROM A SOCIAL SCIENTIST

Christine Bachrach

It's hard to imagine a more "biosocial" research arena than the study of human fertility. Human biology provides the means for reproduction and defines the way in which it occurs, while human actors and the social worlds they occupy exercise considerable control over whether and when they have children. Thus, it is somewhat surprising that this critical interplay of biology and behavior has been so seldom the focus of research. This book takes on the challenge of filling this void. It harnesses the thinking of geneticists and biologists with that of demographers, psychologists, economists, and anthropologists to seek answers about the biosocial processes that influence fertility, answers that cannot emerge from research that focuses on the biological or the social alone. It is a welcome, and important, milestone towards the development of a fully interdisciplinary approach to the science of human fertility.

Is there a gene for fertility? It's a simplistic question, but a serious attempt to answer it opens the door to a maze of fascinating, challenging, and unanswered scientific questions about the role of biological and environmental influences on reproduction. Most scientists would agree that there is no single gene that determines whether a person will have children, or how many, or at what age. They would also agree that our genetic makeup plays an essential role in sexuality and fertility. Genes provide the blueprint for sexual differentiation, reproductive organs, and the endocrine system that regulates reproductive function. Increasingly, scientific evidence points to genetic underpinnings for behavioral predispositions such as sensation-seeking and personality traits such as anxiety. Given the central role of sexual reproduction and the nurturing of young in evolution, it's highly likely that our genetic endowments also contribute to psychological traits and behavioral predispositions that foster the conception, birth, and care of children.

A host of questions are unanswered. In what ways do our genes affect aspects of fertility and fertility behavior that are universal or near-universal in human populations? How much of the individual variation in the volume, timing, and pattern of fertility among human beings can be explained by genetic variation? How can knowledge of the genetic underpinnings of fertility and fertility behavior help to explain aspects of fertility that emerged recently in our evolutionary history, such as contraception and sub-replacement fertility? How do we understand the interplay of genetic and environmental influences in the shaping of individuals' reproductive experience and in the fertility patterns of populations? How can this understanding be used to improve the human condition, for example by fostering understanding of the variety of sexual and reproductive behaviors, improving strategies to discourage sexual and fertility-related behaviors that are not adaptive in today's world, and improving the fit between technologies of fertility management and human needs?

The answers require that attention be brought to bear on a diverse set of processes: from biological processes related to genetic expression to those regulating sexual and reproductive function; from the process and timing of pubertal development to the loss of reproductive function with age; from sexual orientation and sexual behavior to the number, timing, and duration of sexual unions; from fertility motivations, desires, and expectations to childbearing and the nurturant behaviors of parents.

The challenges are significant. The science linking biological influence to behavior in general is not far advanced, although significant strides have been made in some subfields in recent years. Articulating a solid base of knowledge on the pathways through which biological influences can be expected to shape behavior is a critical step towards developing a strong theoretical foundation for biosocial studies. Evolutionary science can offer important insights and a foundation for theory-building. Advances in behavior genetic methodologies are needed to address the complex interplay of genetic and environmental influences. Demographers, sociologists, anthropologists, and economists need to forge working partnerships with psychologists, geneticists, and reproductive biologists. New data will have to be developed that measure potential genetic as well as environmental influences on sexual, fertility, and parenting behaviors.

These questions draw us into challenging ethical issues as well. If study participants provide genetic information to researchers, what information should researchers share with them? Should information be divulged about genetic predispositions for disease, or the biological paternity of children in the family? How can the privacy of genetic information be safeguarded? Suppose a set of genes was found that predisposed individuals to be less monogamous and more promiscuous in their sexual behavior. Could research linking genes and sexual or fertility behavior stigmatize populations who are found to have a higher prevalence of genes for sexual promiscuity? Finally, how can we assure the responsible use of the information produced? The eugenics movements earlier in this century left many in our society wary of genetic research. How do we draw the line between freedom of scientific inquiry and the protection of human rights and disadvantaged populations? Rapidly expanding scientific knowledge generates a constant stream of new ethical challenges that must be addressed on an ongoing basis if we are to advance the scientific agenda responsibly and safely.

Research on the genetic influences on fertility and sexuality is just beginning to blossom. This volume provides a sampling of the questions and approaches that are propelling this field forward into uncharted scientific territory. How do genes affect fertility? The chapters in this volume provide some new answers, and raise a host of new questions that will set the course of research for years to come.

Chapter 1

EVOLUTIONARY CAUSES OF GENETIC VARIATION IN FERTILITY AND OTHER FITNESS COMPONENTS

Kimberly A. Hughes and Mary H. Burleson

Key words: adaptive variation, *Drosophila*, fertility, genetic variation, heritability, longevity, life history traits, maintenance of variation

Abstract: In humans, fertility and other fitness-related traits (life history traits) usually have low to moderate heritabilities, suggesting genetic variability for these traits. This variation is often attributed to some form of "balancing" natural selection that actively maintains it. However, results from evolutionary genetics suggest that most genetic variation is due to non-adaptive forces such as mutation. In this chapter, we apply results of evolutionary quantitative genetics to existing data on human fertility and other fitness components. First, we compare heritabilities and genetic variances for human traits to similar measures in two model organisms: Drosophila *melanogaster* (fruit flies) and *Mus musculus* (house mice), and we argue that genetic variances are much more informative for such comparisons. General results are that genetic variances and heritabilities are similar in humans and in *Drosophila*. The mean coefficient of genetic variation in humans is 11%, and in *Drosophila* 9%. Mean heritability in both species is about 0.2. We also describe both adaptive and non-adaptive mechanisms that can maintain variation within populations, and evaluate evidence relating these mechanisms to human variation. An important implication of this analysis is that only about 12% of phenotypic variation in human fitness traits is likely due to genetic polymorphism that is actively maintained by natural selection. The remainder of phenotypic variance for these traits is either non-genetic, or due to non-adaptive forces such as mutation. We discuss ways in which methods and results of evolutionary quantitative genetics can be applied to studies of human traits.

1. INTRODUCTION

Are humans genetically variable for fertility and fertility-related traits? Several chapters in this volume address this question by describing heritability estimates for traits related to fertility. These estimates provide information about the degree to which individual differences are due to genetic causes rather than to environmental causes. Along with several previously-published reports (Dunne et al., 1997; Herskind et al., 1996; Martin, Eaves, & Eysenck, 1977; McGue, Vaupel, Holm, & Harvald, 1993; Mealey & Segal, 1993; Neel & Schull, 1971; Rodgers, Rowe, & Buster, 1999), these studies support the conclusion that fertility (and other traits related to Darwinian fitness) have low to moderate heritabilities. Measuring heritability in humans is problematic (see below), but assuming that the published estimates are reliable, we can conclude that both genetic and non-genetic factors influence human fertility.

In addition to documenting variation, geneticists are often interested in explaining its origin and persistence. Many recent publications propose adaptive explanations for genetic variation in fertility-related traits (Gangestad & Simpson, 1990; Mealey & Segal, 1993; Segal & McDonald, 1998; Waddington & Youssef, 1996; Wilson, 1994; but see Tooby and Cosmides (1990) for a critique of adaptive explanations). This tendency toward adaptive explanations arises partly from the need for a theoretical framework for behavior genetics (Gangestad & Simpson, 1990; Segal & McDonald, 1998). However, some pressure for authors to generate adaptive hypotheses derives from an assumption common in the literature of evolutionary biology and behavior genetics: that traits closely related to Darwinian fitness are under strong natural selection, and these traits should therefore exhibit little standing genetic variation. Following this logic, substantial heritabilities for fitness traits are assumed to imply that some form of 'balancing selection' must be operating to actively maintain the genetic variance. Otherwise this variance would be effectively eliminated by directional or 'purifying' selection.

Our aim in this chapter is to address the above line of reasoning from the perspective of evolutionary quantitative genetics. This discipline arose soon after the rediscovery of Mendel's principles of genetics (Provine, 1971). Since its inception, a primary goal of evolutionary genetics has been to investigate the causes and consequences of genetic variation in complex (quantitative) traits. Morphological, physiological, and behavioral traits have been investigated, but genetic variation in fertility and other fitness traits has always been of particular interest (Charlesworth & Hughes, 1999; Houle, 1992; Roff, 1992). Consequently, there is a large body of literature on the evolutionary genetics of fitness traits in both natural and laboratory populations.

During the past century, much has been learned about adaptive and non-adaptive explanations for genetic variation in fitness traits. Our goal is to summarize this information and to consider how generalizations drawn from it can be applied to the analysis of human traits. We hope that some of the results from evolutionary genetics can be helpful in interpreting studies of human variation, and perhaps in outlining productive strategies for future investigations.

This chapter is divided into two major sections. In the first, we review heritability estimates for fertility and other human fitness traits and compare them to estimates made in experimental organisms. We also discuss the uses and misuses of heritability as a measure of genetic variation, and describe why evolutionary geneticists now prefer a different metric. The second section describes Fisher's Fundamental Theorem of Natural Selection (the FTNS), and illustrates why moderate levels of genetic variation for fitness traits do not necessarily violate the FTNS. The remainder of that section reviews evolutionary mechanisms that can maintain genetic variance for traits closely tied to Darwinian fitness.

2. COMPARING HERITABILITIES IN HUMANS AND OTHER ORGANISMS

2.1 Fitness Components, Heritabilities, and Genetic Variances

Genetic and environmental influences on quantitative traits are characterized by several measures of variation. For a particular trait, the phenotypic variation within a population, V_P, can be attributed to either genetic or non-genetic sources. Genetic variation can be further partitioned into variation due to different kinds of gene effects and gene interactions, which are described below. Non-genetic variation can be due to age, nutritional status, environmental stressors, learning, social conditions, developmental instability, maternal effects, measurement error, or sampling error. It is usually not possible to determine the exact cause of the non-genetic variation, and all these effects are included in the 'environmental' variance, V_E (Falconer & Mackay, 1996, p. 135).

The heritability of a trait is the proportion of the total phenotypic variation accounted for by genetic variation, V_G. The symbol h^2 usually refers to 'narrow sense heritability,' which is the proportion of V_P accounted for by additive genetic variance, V_A. V_A is the variance due to the average (additive) effects of alleles, and it is the variation that is strictly heritable in the sense that it is responsible for the resemblance between parents and offspring.

The index h^2 does not include non-additive components of genetic variance such as dominance variance, V_D, and epistatic variance, V_I. V_D and V_I are variance terms that arise from dominance interactions between alleles at a locus (V_D) and epistatic interactions between alleles at different loci (V_I). These non-additive terms do contribute to the 'broad-sense heritability,' H^2, which is the proportion of V_P accounted for by all forms of genetic variance. Important relationships to keep in mind are therefore:

$$V_P = V_G + V_E \tag{1}$$

$$V_G = V_A + V_D + V_I \tag{2}$$

$$h^2 = \frac{V_A}{V_P} \tag{3}$$

$$H^2 = \frac{V_G}{V_P} \tag{4}$$

In animals and plants, h^2 is usually estimated from artificial selection experiments or from controlled breeding experiments in which individuals of known relatedness are raised under identical conditions (Falconer & Mackay, 1996). Obviously, neither of these approaches can be used to estimate h^2 for any human trait, so alternate techniques have been devised. The most commonly used methods are twin comparisons, adoption studies and parent-offspring regression (Plomin, Defries, McClearn, & Rutter, 1997).

Methods of estimating human heritabilities rely on more assumptions than do the approaches used in plant and animal research (Falconer & Mackay, 1996, p. 172). For example, twin comparisons ignore non-genetic forces that could cause greater similarity among identical twins than among fraternal twins, and assume that twins are representative of the population as a whole. Also, human studies rely on assumptions about the independence of environments experienced by relatives. While it is clear that environmental similarity among relatives could lead to substantial overestimates of heritability, most empirical studies have suggested that the effect is minor (these are reviewed in Plomin et al., 1997, pp. 73-75; Rowe & Rodgers, 1993). In addition, heritabilities from twin comparisons include both additive and non-additive genetic variance. These estimates are therefore closer to H^2 than to h^2, but are not strictly equivalent to either parameter.

One way to test the validity of human heritability estimates is to compare them with estimates produced for other organisms under controlled environmental conditions. Comparing human heritabilities with those obtained in experimental organisms might therefore be a valuable exercise. Fortunately, evolutionary geneticists have intensively studied heritability and genetic variance for traits (such as fertility) that are closely related to Darwinian fitness.

Some definition of terms may be necessary here. Traits linked directly to Darwinian fitness are known as *fitness components*. Formally, a fitness component is a quantitative trait for which, when all other traits are held constant, an increase in the trait will lead to an increase in total fitness (Charlesworth, 1987; 1994). Fitness itself is measured by r (the intrinsic population growth rate) or by R (the net reproduction rate), which are calculated from age-specific birth and death rates characterizing a particular genotype in a particular population (Charlesworth, 1994). In contrast, morphological traits are not fitness components because they do not have a predictable monotonic relationship to fitness. For example, holding all other traits constant, an increase in fertility at a particular age will lead to an increase in fitness. But an increase in leg length could lead to an increase, a decrease, or no change in total fitness. Fitness components are sometimes referred to as *life-history traits* because they typically relate to maturation, reproduction, or survival. Examples of life-history traits are juvenile survival, development rate, age-specific fecundity, mating success and longevity.

The quantitative genetics of these traits are reviewed by Charlesworth (1987), Mousseau & Roff (1987), Houle (1992), and Charlesworth & Hughes (1999). One of the most important conclusions emerging from these reviews is that, compared to morphological traits, fitness components typically have low heritabilities but moderate to high levels of genetic variance (Houle, 1992). This pattern is somewhat surprising because h^2 is often interpreted as a direct measure of the amount of genetic variation for a trait. However, h^2 is a ratio in which only the numerator is a measure of genetic variation; the denominator is a term that includes both genetic and non-genetic variation. The discrepancy between measures of heritability and genetic variance is due to high levels of non-genetic variation expressed by fitness traits. Traits with high V_E can have low values of h^2, even if they have substantial amounts of genetic variance, because V_E appears in the denominator of the equations specifying h^2 (see Equations 1, 2 and 3).

Why do fitness components have high V_E? Most of the studies reviewed by Houle (1992) were conducted under controlled environmental conditions.

The high V_E values are therefore probably not the result of phenotypic plasticity (meaning that a genotype produces different phenotypes in different environments) or genotype-environment interaction (meaning that environmental variation has different effects on different genotypes). In controlled environments, high V_E values are more likely due to sampling variance, measurement error or developmental instability. Fitness components appear to be quite susceptible to these sources of variance because they are difficult to measure, are often binomially distributed (e.g., mating success and survival), and are influenced by many different biochemical and developmental pathways. Whatever the cause, the observed pattern leads to an important generalization: A large portion of the total phenotypic variance for fitness traits is due to non-genetic effects.

Another valuable insight gained from evolutionary genetics is that heritabilities are not very useful indicators of the absolute amount of genetic variation for a trait. Charlesworth (1987) and Houle (1992) suggested that the coefficient of additive genetic variation, CV_A, is a better metric for comparing variation among traits and among organisms. CV_A is calculated by taking the square root of the additive variance and dividing by the trait mean. It can therefore be thought of as the proportional amount of additive variation relative to the mean (or, when multiplied by 100, it is the variation expressed as a percent of the mean). CV_A is thus a dimensionless metric and is corrected for differences in scale. When Houle compared values among different kinds of traits, he found that fitness components had the highest CV_A values, morphological traits had the lowest, and traits related to growth rate had intermediate values. The pattern observed for h^2 was the reverse. Fitness traits had the lowest values and morphological traits had the highest.

Houle's (1992) result was a very important one because low heritabilities for fitness components had traditionally been taken as evidence that these traits had little genetic variance. This pattern is expected under simple models of selection (e.g., models that ignore mutation and the different forms of balancing selection). But with the realization that traits closely related to fitness actually have *more* genetic variance than other traits, evolutionary biologists began to pay attention to more complex and realistic models. Conclusions from recent analyses of these models are presented in the next section.

2.2   A Comparative Analysis of Variation in Human Fitness Traits

Despite the complications of measuring and interpreting heritabilities, human geneticists are often specifically interested in the amount of phenotypic variation accounted for by genetic variation; ideally, this is what heritability tells us. It is therefore instructive to compare heritabilities for human fitness components to those for analogous traits in other organisms. However, we believe it is even more instructive to compare direct estimates of genetic variation when these are available, because of the properties outlined above.

Table 1 summarizes available estimates of heritability for fitness components in humans. We found twin-study estimates of H^2 for offspring number, longevity, age at first child, and age at first sexual activity. Parent-offspring estimates of h^2 were found only for offspring number and longevity. A few of the studies in Table 1 also reported phenotypic means and standard deviations, allowing us to compute coefficients of genetic variation (CV_A in the case of parent-offspring regressions, and a value close to CV_G for twin studies; here we report both values as percents). Unfortunately, this was only possible for seven of 18 observations reported in Table 1.

Although it is risky to speculate based on only seven values, the CV_G estimate we calculated from Fisher's (1930) data is anomalously high (44.5%) compared to the CV_G values obtained from other studies. This anomaly suggests there may be sources of bias in Fisher's method (a parent-offspring regression) that lead to overestimates of genetic variance. Environmental correlation between parents and offspring is one potential source of bias. Williams & Williams (1974) came to a similar conclusion after a reanalysis of Fisher's data.   We therefore calculated two sets of summary statistics. One set is based on all the values in Table 1, and the other excludes Fisher's estimates.

Heritability estimates for both offspring number and age at first sexual activity are highly variable. In contrast, the five estimates for longevity are very consistent.   We pooled narrow and broad-sense estimates and calculated mean ($\pm SE$) heritabilities for each trait.   Excluding Fisher's estimate, the values are: 0.15±0.05 for offspring number; 0.22±0.06 for longevity; 0.19 (only one estimate) for age at first child; and 0.39±0.06 for age at first sexual activity. Including Fisher's value, mean heritability for offspring number is 0.18±0.05. So, despite variation among studies, mean estimates are roughly similar among different traits, and suggest low to moderate heritabilities for human fitness components.   The overall mean heritability (averaging the trait mean values) is 0.24±0.05 excluding Fisher's estimate and 0.25 ±0.05 including his estimate.

Because we could calculate CV_A or CV_G for so few studies, comparison among traits is probably not meaningful. Averaging within and then across traits, the mean CV_G for human fitness traits is 11.1±0.29 excluding Fisher's data, and 14.7±3.9 including it.

Table 1. Estimates of h^2, H^2, and CV_G for human fitness components

Trait	h^2	H^2	CV_A or CV_G	Analysis Type	Gender	Source[1]
Offspring #	0.40		44.5[2]	Parent-offspring	Female	[1]
"	0.38			Parent-offspring	Female	[2]
"	-0.06		0.0	Parent-offspring	Both	[3]
"		0.19	23.1	Sib analysis	Both	[3]
"	0.18			Parent-offspring	Female	[4]
"	0.14			Parent-offspring	Male	[4]
"		0.06[3]		Twin study	Both	[5]
Longevity		0.23		Twin study	Female	[6]
"		0.26		Twin study	Male	[6]
"		0.22	11.1[2]	Twin study	Both	[7]
"	0.22			Parent-offspring	Female	[8]
"	0.16			Parent-offspring	Male	[8]
Age at first Child		0.19[3]		Twin study	Both	[5]
Age at first sexual activity		0.51	14.3[2]	Twin study	Both	[9]
		0.49/0.32[4]		Twin study	Female	[10]
		0.72/0.00[4]		Twin study	Male	[10]
"		0.15	4.83	Kinship	Female	[11]
"		0.54	12.5	Kinship	Male	[11]

[1]Sources: [1] Fisher 1930, [2] Berent 1953; [3] Imaizumi 1970; [4] Neel & Schull 1971, [5] Mealy & Segal 1993; [6] Herskind et al. 1996; [7] McGue et al. 1993; [8] Westendorp & Kirkwood 1998; [9] Martin et al. 1977; [10] Dunne et al. 1997; [11] Rodgers et al. 1999.

[2]Estimated from means and variances reported in the paper. Value in column labelled CV_G or CV_A is an estimate of CV_A if the data were derived from parent-offspring regressions, but includes non-additive components of genetic variance if derived from twin-studies, sib analysis, or kinship analysis. In the latter case, the estimate is close to the total genetic coefficient of variation CV_G.

[3]Calculated from intraclass correlations of MZ twins reared apart as reported in the paper.

[4]The first number is the value calculated from twins ≤ 40 years of age at the time of data collection. The second value is that calculated for twins > 40 years of age.

To provide a comparison for the human data, we summarized h^2 and CV_A values for fitness traits in the organism most extensively studied with respect to this question, the fruit fly *Drosophila melanogaster* (Table 2). As for humans, different h^2 estimates for the same trait are somewhat variable. However, CV_A estimates for the same trait are very uniform. Even across different traits, mean CV_A estimates are quite consistent (14.5±2.4 for offspring number; 7.8±1.9 for longevity; 10.7±2.4 for male mating success, and 2.5 [only one value] for development time). Averaging trait means, the mean CV_A for fly fitness traits is 8.9±2.0. We can thus state with some confidence that the CV_A for fitness components in a well-characterized model organism is about 9%. The mean h^2 across life-history traits (averaging the mean trait values) is 0.21±0.05.

Table 1. h^2, CV_A, and CV_D estimates (coefficients of dominance variance-see section *3.3*) for fitness components in fruit flies (*Drosophila melanogaster*) and house mice (*Mus musculus*)

Organism	Trait	h^2	CV_A	CV_D	Analysis	Gender	Source[1]
Fruit fly	Offspring #	0.31	17.1	0.3	Sib Analysis	Female	[1]
Fruit fly	"	0.06[2]	11.9[2]		Mixed	Female	[2]
Fruit fly	Longevity	0.03	5.3	17.5	Sib Analysis	Female	[1]
Fruit fly	"	0.11[2]	9.9[2]		Mixed	Both	[2]
Fruit fly	"	0.59	8.3	3.6	Diallel design	Male	[3]
Fruit fly	Mating success[3]	0.08	7.5	0.0	Diallel design	Male	[3]
Fruit fly	Mating success[4]	0.18	13.9	0.0	Diallel design	Male	[3]
Fruit fly	Develop. time	0.28[2]	2.5[2]		Mixed	Both	[2]
Mouse	Litter size	0.37			Sib Analysis	Female	[4]
Mouse	"	0.20	12.6	2.8	Selection	Female	[5]
Mouse	"	0.24[5]	19.1		Selection	Female	[5]
Mouse	Repro. life	0.11	21.9		Selection	Both	[6]

[1]Sources: [1] Rose & Charlesworth, 1981; [2] Houle, 1992; [3] Hughes, 1995; [4] Falconer, 1963; [5] Falconer, 1965; [6] Nagai et al., 1995.
[2]Median values from several studies were calculated by Houle (1992). Number of studies from which median estimates were calculated: fecundity: 12; longevity: 7; development time: 2. Houle did not calculate CV_D values.
[3]Competitive male mating ability measured at 3 days of age.
[4]Competitive male mating ability measured at 21 days of age.
[5]Calculated from mean of two heritability estimates reported in the paper.

We also found a few published estimates of heritabilities in the house mouse, *Mus musculus* (Table 2). The mean h^2 across traits is 0.19±0.07. We could calculate CV_A values from the data in two of these studies; the mean CV_A is 18.9±2.8. This value is slightly higher than those for either humans or flies. Mouse studies may yield somewhat higher estimates because they generally are performed on hybrid populations. These hybrids are produced from crosses between inbred lines or crosses between different natural populations, and are therefore far from evolutionary equilibrium. They do provide the only comparable data from a non-human mammal, however. Consequently, it is worth noting that the human estimates are at least as consistent with the fly estimates as with those from a mammalian model organism.

Analysis of variance shows that heritability for human life history traits is not statistically different from that for flies or for mice, even with Fisher's estimate included (human vs. fly: $F_{1,28} = 0.37$, $P > 0.5$; human vs. mouse: $F_{1,22} = 0.12$, $P > 0.50$). The same holds for coefficients of variation (human vs. fly: $F_{1,13} = 1.30$, $P > 0.25$; human vs. mouse: $F_{1,8} = 0.06$, $P > 0.50$).

In a recent comparative analysis, Burt (1995) combined quantitative-genetic, mutation-rate, and gene-flow data to arrive at estimates of the rate of change of fitness in natural and laboratory populations. This rate is directly proportional to the genetic variance for fitness, and so can be used to estimate CV_A for fitness itself. He used data from flowering plants (5 species), rotifers, *Drosophila*, birds (2 species), and humans to conclude that CV_A for fitness fell between 10 and 30%. This range brackets most of the values reported for human fitness traits, and provides independent confirmation that a mean CV_G of 10-15% is reasonable.

Inspection of Tables 1 and 2 suggests that heritability estimates are more variable within species than are coefficients of variation. So it is somewhat surprising that the mean h^2 (or H^2) estimates for all three species are so similar. However, heritabilities do not reliably predict the magnitude of genetic variation. Combining the estimates from all the studies in Tables 1 and 2, the correlation between heritability and coefficients of variation is not significant (Pearson's $r = 0.21$, $P > 0.40$; Spearman's $\rho = 0.33$, $P > 0.20$ with Fisher's data excluded; Pearson's $r = -0.31$, $P > 0.20$; Spearman's $\rho = 0.38$, $P > 0.10$ with Fisher's data included). This case thus provides another example of h^2 values giving a misleading picture if they are interpreted as comparisons of genetic variability.

2.3   Section Summary

Heritability values for human fitness traits are consistent with estimates from other organisms. However, heritability does not provide a reliable indication of genetic variation within a population. When we compare genetic coefficients of variation (CV_A or CV_G), we find that human populations apparently contain levels of variation similar to that seen in experimental organisms. In particular, the variance is only slightly (and non-significantly) higher than that reported for fitness components in *D. melanogaster*. This conclusion is based on only a few available estimates from the human genetics literature. More estimates from human studies are therefore needed to confirm our preliminary comparison. As we discuss in the next section, genetic variances are also more useful than heritabilities when the goal is to test different evolutionary models. For both these reasons, we suggest that human geneticists report coefficients of variation along with heritabilities in their studies of fitness traits. As more of these estimates become available, comparisons with other organisms will become more reliable, and quantitative tests of evolutionary models will become possible.

3.　　SIX SOURCES OF GENETIC VARIATION

If heritability estimates for human fertility and other fitness traits can be taken at face value, they suggest that human populations contain levels of genetic variation similar to that seen in other organisms. In this section, we describe various evolutionary forces that can account for this variation.

This variation requires an explanation because of a mathematical formula first developed by R.A. Fisher and known as Fisher's Fundamental Theorem of Natural Selection (the FTNS; Fisher, 1930). The FTNS states that the rate of evolutionary increase in the mean fitness of a population is equal to the amount of additive genetic variance for fitness within the population. Extensions and generalizations of the theorem (Kimura, 1958) predict that populations near evolutionary 'equilibrium' will exhibit little or no additive genetic variance for fitness.

The FTNS is often interpreted to mean that traits like fecundity, fertility, mating success, and longevity should exhibit little or no additive variation within populations. These traits are subject to very strong natural selection because they largely determine organismal fitness. Consequently, simple models of natural selection predict that genetic variation will be eroded very quickly.

Below, we review several evolutionary forces that can maintain genetic variation. These forces are potential solutions to the difficulties posed by Fisher's Theorem. We briefly describe mutation, frequency-dependent selection, heterozygote advantage, environmental heterogeneity, sexual antagonism, and antagonistic pleiotropy. Gene flow between different populations can also maintain variation (Crow & Kimura, 1970), but here we will consider only those factors acting within single populations.

For some well-studied organisms, we now have enough information to infer which of these factors contributes most to phenotypic and genetic variation. Humans may experience quantitatively different levels of mutation or selection than do other species, but the qualitative properties of these forces are universal. Given comparable data, human variation should be amenable to the same analyses that are used in other organisms. Consequently, data and conclusions from other species can guide us in the search for causes of human variation. Therefore, we also review what is known about the maintenance of variation in other organisms, and relate these results to comparable human data.

3.1 Mutation-Selection Balance

Mutation is the most ubiquitous of the phenomena that maintain genetic variance in fitness. It is worth noting that Fisher's FTNS explicitly ignores the effects of mutation. Fisher himself describes the maintenance of variation by mutation later in the same volume in which he proposes the Fundamental Theorem (Fisher, 1930).

For some organisms, we know a good deal about effects of mutations on fitness components, and about the amount of variation that can be attributed to this force. *Drosophila* fitness traits have been the object of mutation experiments and quantitative genetic studies for over 40 years. This body of literature provides the most complete picture available on the genetic variation for fitness and its components in a single species (reviewed in Charlesworth & Hughes, 1999). Information from mutation experiments can be used to estimate mutation rates for alleles affecting fitness and its components. We can then compare the amount of existing variation to the amount expected given the observed mutation rate. This comparison provides a test of a 'null model' of evolutionary genetics, the model suggesting that genetic variation is maintained by a balance between mutation and the removal of mutations by natural selection. This hypothesis is known as the *mutation-selection balance* model.

When we compare mutation rates in *Drosophila* to standing variation, we find that mutation cannot account for all of the observed additive variation for fitness components, but it can account for a substantial portion of it (reviewed in Charlesworth & Hughes, 1999). For example, mutation alone can account for between one-third and two-thirds of the V_A for traits like female fecundity, male mating success and juvenile survival. Thus, a very large proportion of within-population variation is due to mutation. Similar conclusions have been drawn from data on other organisms (Houle, Morikawa, & Lynch, 1996; Lynch et al., 1998). Further, the vast majority of these mutations are deleterious (Elena et al., 1998; Fisher, 1930), and the variation is explicitly non-adaptive (Lande, 1995; Lynch, Conery, & Burger, 1995).

Unfortunately, we cannot make a similar calculation for human fitness traits. Although we do have information on standing variation (see Table 1), we do not have information on mutational effects (mutation rate, average effect of mutations, and average dominance of mutations) that would allow us to make the comparison for our own species (but see Eyre-Walker & Keightley, 1999, for an attempt to measure the deleterious mutation rate in hominids). Nevertheless it is probably safe to assume that the mutation rate for human fitness traits is at least as high as the rate in *Drosophila* (Crow, 1993).

In fruit flies, some of the genetic variation that cannot be explained by mutation may be maintained by one of the forms of 'balancing selection' described below. But before describing these mechanisms, we would like to emphasize that a good deal of the unexplained variance in *Drosophila* could be caused by experimental sources of bias. For example, during an experiment the differences between the environment to which a population is adapted and the environment actually experienced can lead to inflation of the genetic variance. This is called a 'novel-environment effect.' The effect occurs because the variance contributed by a polymorphic locus is directly proportional to the average effect of allele substitutions at the locus. In a novel environment, the effects of alleles are likely to be different from those in the original environment. Moreover, allelic effects will tend to be larger in a novel environment since, under mutation-selection balance, alleles with large effects in the original environment will either have been eliminated or have gone to fixation. This effect has been demonstrated in experiments by measuring genetic variance and the effects of individual loci in different environments (Fry, et al., 1998; Kondrashov & Houle, 1994). The potential for novel-environment effects cannot be completely eliminated even by using populations adapted to laboratory conditions. Even if experimental conditions are only slightly different from normal culture conditions, inflation of genetic variance can occur.

Clearly, variation due to mutation and novel-environment effects cannot be considered 'adaptive variation.' We can consequently draw two important conclusions from the *Drosophila* data. The first is that most phenotypic variation for fitness traits is not genetic. This conclusion derives from heritability values, which average about 20% (see Section 2.2). The second conclusion is that most genetic variation cannot be considered adaptive because it is due to non-adaptive forces (primarily mutation). Therefore, if the same forces acting on other organisms shape human variation, we should be extremely cautious in attributing adaptive significance to the variation itself.

So how much human variation *is* likely to be due to selective maintenance of genetic polymorphism? Remember that broad-sense heritabilities for human fitness traits average 24% (see Section 2.2). This implies that 24% of the phenotypic variation is genetic. Also remember that, at least in *Drosophila*, one- to two-thirds of all genetic variation is mutational variance, and some of the remainder is probably due to experimental bias. So, assuming about half of human genetic variation is mutational, *the upper limit to the amount of phenotypic variation due to adaptive genetic polymorphism is 12%. The remaining 88% of within-population variation is due to deleterious mutation, developmental instability, and environmentally-induced variation (phenotypic plasticity) that may or may not be adaptive* (Kingsolver & Huey, 1998; Scheiner, 1993; Winn, 1998).

Having duly given warning that most phenotypic variation probably cannot be explained by selective forces, we proceed to review the various forms of balancing selection that are capable of maintaining genetic polymorphism. Where possible, we have included examples of human conditions attributed to the different mechanisms. In both *Drosophila* and humans, some genetic variance is maintained by selection. Clearly, a few polymorphisms are selectively maintained (e.g., human leukocyte antigen [HLA] polymorphism and alcohol dehydrogenase [Adh] polymorphism in *Drosophila*). Nevertheless, it is usually very difficult to determine which form of selection is operating, even when we are dealing with experimental organisms. The extensive literature on Major Histocompatibility Complex [MHC] and HLA polymorphism in vertebrates (Hedrick, 1994), and Adh polymorphism in *Drosophila* (Van Delden, 1997) illustrates this difficulty. With few exceptions, evidence relating human variation to specific selective mechanisms remains inconclusive.

3.2   Frequency-Dependent Selection

Frequency-dependent selection occurs when the fitness of a genotype is inversely proportional to its frequency in the population (Crow & Kimura 1970, pp. 256-257). There are no clear-cut cases of the maintenance of human polymorphism by this type of selection. The extreme polymorphism of HLA genes is most simply explained by frequency dependence, but this and other hypotheses for HLA diversity have been very controversial (Hedrick, 1994). HLA genes encode critical components of the immune system in humans and in all other vertebrates (where they are referred to as MHC genes). They are extremely polymorphic in all organisms in which they have been found, so whatever force is responsible for maintaining the diversity must be widespread.

In addition to single-gene polymorphisms, morphological and behavioral variation in other vertebrates has been attributed to frequency-dependent selection. For example, male guppies (*Poecilia reticulata*) are highly polymorphic for genetically-determined color patterns. Hughes et al. (1999) reported that males with novel patterns had higher reproductive success than males with common patterns, and that the difference was due to female preference for novelty. This result is consistent with frequency-dependent mate choice being responsible for color polymorphism in this species.

Frequency-dependent mating success is also believed to operate in ruffs, *Philomachus pugnax* (Lank, Smith, Hanotte, Burke, & Cooke, 1995). Male birds display genetic polymorphism for breeding plumage and reproductive behavior. Behaviorally, males are either 'independents' that defend mating arenas (courts), or are nonterritorial 'satellites' that move among courts. Satellites are recruited by independents to temporarily share a court (van Rhijn, 1991). Independents benefit from sharing their courts with satellites because females prefer mating with males on co-occupied courts; satellites benefit because they usually mate only if they are sharing a court with an independent (Hugie & Lank, 1996). Frequency dependence occurs because a male of either type experiences a higher probability of mating when his type is rare than when his type is common. This phenomenon can account for maintenance of both male types within populations.

3.3   Heterozygote Advantage

In principle, heterozygote advantage (overdominance) can maintain genetic variation for fitness or its components. If the heterozygote at a locus has higher fitness than either homozygote, the total genetic variance in fitness generated by the locus is equal to V_D, the dominance variance (Haldane, 1949). So heterozygote advantage produces no V_A at all. A testable prediction of the model is that we should see large amounts of V_D relative to V_A if heterozygote advantage maintains substantial genetic variance. On the other hand, if heterozygote advantage is unimportant in maintaining variation, then V_D should be small compared to V_A. Although Haldane's original conclusion was based on a simple one-locus model, his result has since been extended to characters controlled by many loci (Schnell & Cockerham, 1992).

Based on comparisons of V_A and V_D in *Drosophila*, heterozygote advantage does not seem generally important in the maintenance of variation for fitness traits (see CV_D estimates in Table 2). V_A has been substantially larger than V_D in almost all studies of egg-to-adult viability (Mukai, 1985), female fecundity (Rose, 1981), male mating success (Hughes, 1995), and male longevity (Hughes, 1995). For two traits, female longevity (Rose,

1981) and sperm competitive ability (Hughes, 1997), there is some evidence that V_D is larger than V_A. For these two traits, a few loci with strongly overdominant effects may account for most of the genetic variation.

Comparison of V_A and V_D thus gives us one way to find variation that is maintained by selection. If estimates of the inbreeding load (B) are available, we can go even farther. We can use V_D and B to estimate the minimum number of segregating loci affecting a trait, and the average effect of these loci (Charlesworth & Hughes, 1999; Mukai et al., 1974). For one *Drosophila* trait, sperm competitive ability, this technique suggested that a few loci with large effects were contributing most of the variation (Hughes, 1997).

In humans, it is clear that at least one single-gene polymorphism is maintained by heterozygote advantage. This is the textbook case of polymorphism for sickle cell anemia in populations where malaria is endemic (Hartl & Clark, 1997, pp. 230-231). There is some evidence for a similar effect associated with cystic fibrosis (Gabriel, Brigman, Koller, Boucher, & Stutts, 1994; Schroeder, Gaughan, & Swift, 1995). Still, clear examples of overdominant alleles are very few, despite the large amount of effort that has been devoted to testing this hypothesis.

3.4   Spatial and Temporal Variation in the Environment

It seems intuitively obvious that spatial or temporal variation in environmental conditions will maintain genetic variation. However, mathematical models show that variable environments will maintain genetic variation only under fairly restrictive conditions (Hedrick, 1986; Hoekstra, Bijlsma, & Dolman, 1985). Maintenance of polymorphism requires either strong selection on single loci, or a form of mean heterozygote advantage such that (averaged over all environments) heterozygotes have higher fitness than homozygotes (Hoekstra et al., 1985; Maynard Smith & Hoekstra, 1980). Hedrick (1986; see also Hedrick, Ginevan, & Ewing, 1976) considered both spatial and temporal variation in the environment, and concluded that although conditions for both cases are stringent, spatial variation is more likely to maintain polymorphism than is temporal variation. Habitat selection (Hedrick, 1990; Hoekstra et al., 1985), and limited gene flow among different environments (Christiansen, 1975) make the conditions somewhat less restrictive.

There has been continuing progress in theoretical treatment of this mechanism (Gillespie, 1991; Gillespie & Turelli, 1989). However, there have been only a few empirical tests relating quantitative variation to environmental variation. In *D. melanogaster*, Mackay (1980; 1981) observed increased V_A and h^2 for three morphological traits when she varied the concentration of alcohol spatially and temporally. Hedrick (1986) has criticized these results on methodological grounds, citing lack of appropriate controls, and pointing out several results of the experiment that are inconsistent with predictions of the model. In similar experiments, R.A. Riddle and co-workers also obtained equivocal results; they were unable to document a consistent association between genetic variation and variation in culture conditions in two species of flour beetles (Dawson & Riddle, 1983; Riddle, Dawson, & Zirkle, 1986; Zirkle & Riddle, 1983). So despite the intuitive appeal of this mechanism, its general importance in maintaining quantitative variation is uncertain.

3.5   Sexual Antagonism

If an allele that increases fitness in one sex causes a decrease in fitness in the other, then it is said to have sexually-antagonistic effects. Sexual antagonism can maintain genetic variation, and Haldane (1962) and Livingstone (1992) have described the conditions for the two-allele case. In general, sexual antagonism can lead to the maintenance of variance in two ways: by causing the fitness of the heterozygote to be higher than either homozygote when averaged over sexes, or by generating rather large fitness differences between the sexes. Although this phenomenon has received little attention compared to other mechanisms for maintaining variation, several recent *Drosophila* studies indicate that populations are polymorphic for sexually antagonistic alleles and that new mutations can have different effects in males and females (Mackay & Fry, 1996; Mackay, Fry, Lyman, & Nuzhdin, 1994; Mackay, Lyman, & Hill, 1995; Mackay, Lyman, Jackson, Terzian, & Hill, 1992; Rice, 1992; Rice, 1996).

Livingstone (1992) used the concept of sexual antagonism to model the maintenance of human hereditary hemochromatosis. The prevalence of hemochromatosis is over 10% in some human populations and the disease affects men much more severely than women. The proposed advantageous effect of the allele is that women may gain protection from iron-deficiency anemia. Livingstone concluded that the level of polymorphism is consistent with a sexually antagonistic model. In order for the polymorphism to be maintained, however, the selective advantage accruing to females must be nearly exactly balanced by the disadvantage accruing to males. There is apparently no direct evidence that the selection coefficients meet these conditions.

3.6　　Genetic Correlations

Fisher's Fundamental Theorem and Kimura's generalization are statements about total fitness. High additive variances for fitness components are not necessarily inconsistent with low additive variance for total fitness. Genetic "tradeoffs" between different traits can lead to substantial genetic variance for fitness components, but little variation for fitness itself.

A genetic trade-off is defined as a negative genetic covariance (or correlation) between different traits. These tradeoffs can be due to individual alleles that have effects on more than one fitness component (pleiotropy) or to linkage disequilibrium among alleles affecting different traits. Linkage disequilibrium occurs when alleles at different loci are inherited non-independently (i.e., particular alleles are inherited together more often than expected under Mendel's law of independent assortment, usually because they are located close together on the same chromosome). An example of a trade-off due to pleiotropy would occur if there were two alleles at a locus, and one allele caused increased fecundity but decreased longevity compared to the other. These two alleles will then contribute to a negative genetic correlation between fecundity and longevity. Alternately, a negative correlation can result if an allele causing high fecundity is in linkage disequilibrium with an allele causing lowered longevity.

Negative genetic correlations between fitness components allow a population to obey the Fundamental Theorem, even if some of the traits have large amounts of additive genetic variance. This concept is best explained algebraically. If fitness is controlled by many loci with small effects on different fitness traits, the additive genetic variance in total fitness is given approximately by the sum of the genetic variances for each trait, plus the additive genetic covariances between each pair of traits:

$$V_A = \sum_i \sum_j g_{ij} \left(\frac{\partial w}{\partial z_i} \right) \left(\frac{\partial w}{z_j} \right). \tag{2}$$

The summation is carried out over the different components of fitness. The i and j subscripts refer to specific traits (e.g., fecundity, longevity, and male mating success), z_i stands for the different values that a specific trait can assume (e.g., one, two, or three offspring for the fecundity trait), and g_{ij} is the additive genetic covariance between traits (Charlesworth, 1984). Since the additive genetic variance of a trait is the same as the additive covariance of the trait with itself, these enter the summation when $i = j$. The other terms in the equation describe the change in total fitness caused by a small change in a particular fitness component. These are the partial derivatives of fitness with respect to its components, and are denoted as $\partial w/\partial z_i$ and $\partial w/\partial z_j$. The partial derivatives are positive by definition (see the definition of a fitness component given in the previous section) and the genetic variances also must be positive. Therefore, there is only one way that V_A for total fitness can be small when additive variances for individual components (g_{ii}) are large: when at least some of the genetic covariances (g_{ij}) are negative (Dickerson, 1955; Rose, 1982). This result leads to the main prediction of this model: there should be negative genetic correlations between some fitness components.

3.7 Maintenance of Variation has Practical Implications: Case Study of the Genetics of Aging

Although aging and senescence are not the primary focus of this volume, we hope this section will illustrate some practical implications of understanding the maintenance of variation. What follows is a brief review of an extensive literature on the evolutionary genetics of senescence. These studies have focused on understanding the genetic basis of senescence and the evolutionary processes that have led to this near-universal biological pattern.

Two evolutionary models can explain the phenomenon of senescence, and the models rely on different assumptions about the maintenance of variation. The *antagonistic-pleiotropy* theory of senescence postulates that genetic correlations between fitness traits (see section 3.6) are the underlying cause of senescence. These genetic correlations arise because individual genes affect more than one trait (pleiotropy). It is these pleiotropic genes that cause health and performance to decrease with advancing age (Rose, 1991; Williams, 1957).

To understand the theory, first assume that some alleles have pleiotropic effects that are limited to certain ages (are age-specific in expression). If an allele has beneficial effects at one age, but deleterious effects at a different age, it is subject to conflicting selection pressures. An example of such an allele would be one that boosts production of a hormone that increases fertility during early adulthood, but also increases risk of disease later in life. The strength of selection acting on this allele is inversely proportional to the age of expression. This is because late-life increases or decreases in reproduction and survival have little effect on Darwinian fitness compared to similar effects early in life (Charlesworth, 1994). Natural selection will thus tend to favor alleles with beneficial early-age effects and deleterious late-age effects. Conversely, alleles with the reverse pattern of action (early deleterious effects and late beneficial effects) will be selected against. As the frequency of the first type of allele increases in the population and the frequency of the second type decreases, the population will evolve a senescent life history. Individual differences in longevity can be caused by polymorphism maintained by these genetic 'trade-offs' (Rose, 1985).

A competing theory of the evolution of senescence is based on a model of mutation-selection balance (see section 3.1). This model assumes that some alleles will have unconditionally deleterious effects, but the effects will be confined to certain ages. One potential example of this kind of allele is the one causing Huntington's Disease (HD) in humans: people with the HD allele usually have no symptoms until late middle age. The premise of the *mutation-accumulation model* is that natural selection will be efficient at eliminating deleterious mutations if their effects are expressed early in life, but much less efficient if the effects are only expressed late in life. Again, this is because the strength of selection declines with age. Thus, the population-wide frequency of a mutant allele (determined by the balance between mutation and selection) will be higher the later the age of expression (Charlesworth, 1994; Medawar, 1952). High frequency of deleterious alleles that are only expressed at late ages will cause performance to decrease with advancing age, and will contribute to individual longevity differences.

Both the antagonistic-pleiotropy and mutation-accumulation models have been tested several times (most extensively in *D. melanogaster*). Partridge and Barton (1993) and Charlesworth (1994) provide recent reviews. The general conclusion emerging from these reviews is that both mechanisms probably contribute to senescence in fruit flies.

Directly testing these models in humans will be difficult. However, new gene-mapping techniques that use extended human pedigrees provide an opportunity to detect effects of individual loci. So it is likely that examples of genes falling into the two categories will accumulate, although determining which class of genes is the most important contributor to human senescence may not yet be feasible.

3.8   Section Summary

In both humans and fruit flies, heritabilities of about 0.2 for fitness traits suggest that about 20% of the phenotypic variation for these traits has a genetic basis. Several different mechanisms are potentially responsible for the maintenance of this genetic variation. The most ubiquitous of these is mutation. Mutation-selection balance is responsible for between one-third and two-thirds of the genetic variance for fitness traits in *Drosophila*. The remainder of the genetic variance could be due to novel-environment effects or to one of several forms of balancing selection. It is currently unknown, even for experimental organisms, which of these mechanisms contributes most to the variation that is not due to mutation. However, heterozygote advantage (overdominance) can probably be ruled out as a generally important mechanism. Practical results of studying the maintenance of variation include gaining insight into genetic and physiological processes (e.g., the functioning of the vertebrate immune system) and elucidating determinants of genetic-disease prevalence in different populations. Also, an understanding of the causes of variation has been critical in uncovering the genetic and evolutionary mechanisms underlying senescence.

4. CONCLUSIONS

Many discussions of human behavioral and reproductive variation have emphasized the adaptive nature of the variation, whether it is due to genetic causes or to the ability of organisms to produce different phenotypes in response to different environments (Gangestad & Simpson, 1990; Mealey & Segal, 1993; Segal & McDonald, 1998; Tooby & Cosmides, 1990; Waddington & Youssef, 1996; Wilson, 1994). While some phenotypic variation is undoubtedly adaptive, it is likely that only a small proportion is based on genetic polymorphism that is actively maintained by selection. Thus, an over-emphasis on the adaptiveness of human variation ignores some of the fundamental tenets of evolutionary genetics.

Evolutionary geneticists have long recognized that many forces contribute to *non-adaptive* variation. Mutation is the ultimate source of genetic variation, and is thought to be responsible for a very substantial fraction of genetic variation within populations (see Section 2). Between-population variation can also be non-adaptive. For example, statistical sampling effects (genetic drift) are known to be an important source of between-population variation. In fact, genetic studies of populations usually begin with the assumption that drift is responsible for any differences seen. It is only if this null hypothesis is rejected that adaptive scenarios are proposed to explain population differences (Lynch, 1994).

Two primary causes of genetic variation are therefore explicitly non-adaptive. Consequently, results from both theoretical and experimental genetics caution against automatically seeking adaptive explanations for genetic variation for human traits. This caution should extend even to fertility and other traits closely tied to Darwinian fitness.

On the other hand, evidence from non-human animals indicates that some genetic variation is adaptive. If studies of experimental organisms can be used as a guide, up to 12% of the phenotypic variation in human fertility, longevity, and mating success could be adaptive genetic variation that is maintained by some form of balancing selection. The remainder of the phenotypic variation for these traits is likely to be either non-genetic or due to non-adaptive sources of genetic variation such as mutation.

It is more difficult to put a numerical estimate on the amount of non-genetic variation that is likely to be adaptive. Clearly, organisms can produce different phenotypes in response to environmental variation. This non-genetic response to environmental variation is known as *phenotypic plasticity*. To our knowledge, evolutionary biologists have not yet attempted to quantify relative proportions of adaptive vs. non-adaptive phenotypic plasticity in any species. So far, most studies of plasticity in plants and animals have been directed at variation in traits that are especially likely to be adaptive. For example, a large number of animal studies have been conducted on species that exhibit two or three discrete morphological classes (Parejko & Dodson, 1991, Kingsolver, 1998). Moreover, these studies have usually focused on fairly drastic variation in external morphology (such as the presence or absence of spines or other structures). Even with these inherent biases, both adaptive and non-adaptive plasticity has been documented (Dudley & Schmitt, 1996; Kingsolver & Huey, 1998; Scheiner, 1993; Winn, 1998). The occurrence of several negative results suggests that non-adaptive plasticity may be quite common. This question will hopefully be addressed in future studies of plasticity.

In conclusion, there are obviously many avenues for future work on human variation in fertility and other fitness traits. With respect to the issues raised in this chapter, we would like to suggest a few directions that might be fruitful. First, as more estimates of additive and non-additive variance become available for human traits, the heterozygote-advantage model can be tested. In particular, values of CV_D that are large relative to CV_A provide indirect evidence for variation that is maintained by this form of balancing selection.

Second, there are methods available for estimating the number of polymorphic genes affecting a trait. These techniques require values of V_D together with estimates of the effects of inbreeding. Application of these methods to human data can reveal whether different traits have different genetic "architecture", and they can provide preliminary indication of genes of major effect. If evidence for major-effect genes is found for a particular trait, that trait becomes a good candidate for genetic mapping studies. Consequently, this approach could facilitate progression of behavior genetic studies to the stage of identifying genes affecting traits of interest.

Finally, the 80% of human life-history variation that is non-genetic is clearly worthy of investigation. Evolutionary biologists are beginning to formalize methods to test whether or not phenotypic plasticity is adaptive (Kingsolver & Huey, 1998; Komers, 1996). Some of these methods may be applicable to data from human studies, although some clearly will not be (e.g., replicating genotypes in many different environments). We believe that determining which of these tests are applicable, and using human data to test the adaptive plasticity hypothesis will be a challenging and rewarding endeavor for students of human fertility and human behavior.

ACKNOWLEDGEMENTS

We thank K. Dixon, P. Hedrick, J. Rodgers, and B. Sullivan for providing very helpful comments on drafts of this chapter. K.A.H. was supported by National Science Foundation grant DEB-9734008 during preparation of the manuscript, and M.H.B. was supported by Arizona Disease Control Research Commission Contract #9906.

REFERENCES

Burt, A. (1995). The evolution of fitness. *Evolution, 49*, 1-8.
Charlesworth, B. (1984). The evolutionary genetics of life-histories. In B. Shorrocks (Ed.), *Evolutionary Ecology* (pp. 117-133). Oxford, U.K.: Blackwell.
Charlesworth, B. (1987). The heritability of fitness. In J. W. Bradbury & M. B. Andersson (Eds.), *Sexual Selection: Testing the Alternatives* (pp. 21-40). Chichester, U.K.: John Wiley.
Charlesworth, B. (1994). *Evolution in Age-Structured Populations*. (2nd ed.). Cambridge, U.K.: Cambridge University Press.
Charlesworth, B., & Hughes, K. A. (1999). The maintenance of genetic variation in life history traits. In R. S. Singh & C. B. Krimbas (Eds.), *Evolutionary Genetics from Molecules to Morphology* (Vol. 1, pp. 369-391). Cambridge, UK: Cambridge University Press.
Christiansen, F. B. (1975). Hard and soft selection in a subdivided population. *Ameican Naturalist, 109*, 11-16.
Crow, J. F. (1993). How much do we know about spontaneous human mutation rates? *Environmental and Molecular Mutagenesis, 21*, 122-129.
Crow, J. F., & Kimura, M. (1970). *An Introduction to Population Genetics Theory*. New York: Harper and Row.
Dawson, P. S., & Riddle, R. A. (1983). Genetic variation, environmental heterogeneity, and evolutionary stability. In C. F. King & P. S. Dawson (Eds.), *Population Biology: Retrospect and Prospect* . New York: Columbia University Press.
Dickerson, G. E. (1955). Genetic slippage in response to selection. *Cold Spring Spring Harbor Symp. Quant. Biol., 20*, 213-224.

Dudley, S. A., & Schmitt, J. (1996). Testing the adaptive plasticity hypothesis: Density dependent selection on manipulated stem length in *Impatiens capensus*. *American Naturalist, 147*, 445-465.

Dunne, M. P., Martin, N. G., Statham, D. J., Slutske, W. S., Dinwiddie, S. H., Bucholz, K. K., Madden, P. A. F., & Heath, A. C. (1997). Genetic and environmental contributions to variance in age at first sexual intercourse. *Psychological Science, 8*, 211-216.

Elena, S. F., Ekunwe, L., Hajela, N., Oden, S. A., & Lenski, R. E. (1998). Distribution of fitness effects caused by random insertion mutations in Escherichia coli. *Genetica (Dordrecht), 102-103*, 349-358.

Eyre-Walker, A., & Keightley, P. D. (1999). High genomic deleterious mutation rates in hominids. *Nature, 397*, 344-347.

Falconer, D. S., & Mackay, T. F. C. (1996). *Introduction to Quantitative Genetics*. (4 ed.). Essex, U.K.: Longman.

Fisher, R. A. (1930). *The Genetical Theory of Natural Selection*. Oxford: Oxford University Press.

Fry, J. D., Nuzhdin, S. V., Pasyukova, E. G., & Mackay, T. F. C. (1998). QTL mapping of genotype-environment interaction for fitness in *Drosophila melanogaster*. *Genetical Research, 71*, 133-141.

Gabriel, S. E., Brigman, K. N., Koller, B. H., Boucher, R. C., & Stutts, M. J. (1994). Cystic fibrosis heterozygote resistance to cholera toxin in the cystic fibrosis mouse model. *Science, 266*, 107-109.

Gangestad, S. W., & Simpson, J. A. (1990). Toward an evolutionary history of female sociosexual variation. *Journal of Personality, 58*, 69-92.

Gillespie, J. H. (1991). *The Causes of Molecular Evolution*. Oxford: Oxford University Press.

Gillespie, J. H., & Turelli, M. (1989). Genotype-environment interactions and the maintenance of polygenic variation. *Genetics, 121*, 129-138.

Haldane, J. B. S. (1949). Parental and fraternal correlations in fitness. *Annals of Eugenics, 14*, 288-292.

Haldane, J. B. S. (1962). Conditions for stable polymorphism at an autosomal locus. *Nature, 193*, 1108.

Hartl, D. L., & Clark, A. G. (1997). *Principles of Population Genetics*. (3rd ed.). Sunderland, MA: Sinauer Associates.

Hedrick, P. W. (1986). Genetic polymorphism in heterogeneous environments: A decade later. *Annual Review of Ecology and Systematics, 17*, 535-566.

Hedrick, P. W. (1990). Genotype-specific habitat selection: A new model. *Heredity, 65*, 145-149.

Hedrick, P. W. (1994). Evolutionary genetics of the major histocompatibility complex. *American Naturalist, 143*, 945-964.

Hedrick, P. W., Ginevan, M. E., & Ewing, E. P. (1976). Genetic polymorphism in heterogenous environments. *Annual Review of Ecology and Systematics, 7*, 1-32.

Herskind, A. M., McGue, M., Holm, N. V., Sorensen, T. I. A., Harvald, B., & Vaupel, J. W. (1996). The heritability of human longevity: A population-based study of 2872 Danish twin pairs born 1870-1900. *Human Genetics, 97*, 319-323.

Hoekstra, R. F., Bijlsma, R., & Dolman, A. J. (1985). Polymorphism from environmental heterogeneity: Modles are only robust if the heterozygote is close in fitness to the favoured homozygote in each environment. *Genetical Research, 45*, 299-314.

Houle, D. (1992). Comparing evolvability and variability of quantitative traits. *Genetics, 130*, 195-204.

Houle, D., Morikawa, B., & Lynch, M. (1996). Comparing mutational variabilities. *Genetics, 143*, 1467-1483.

Hughes, K. A. (1995). The evolutionary genetics of male life-history traits in *Drosophila melanogaster. Evolution, 49*(5), 521-537.

Hughes, K. A. (1997). Quantitative genetics of sperm precedence. *Genetics, 145*(1), 139-151.

Hughes, K. A., Du, L., Rodd, F. H., & Reznick, D. N. (1999). Familiarity leads to female mate preference for novel males in the guppy, *Poecilia reticula. Animal Behaviour, In Press.*

Hugie, D. M., & Lank, D. B. (1996). The resident's dilemma: A female choice model for the evolution of alternative mating strategies in lekking male ruffs (*Philomachus pugnax*). *Behavioral Ecology, 8*, 218-225.

Kimura, M. (1958). On the change of population fitness by natural selection. *Heredity, 12*, 145-167.

Kingsolver, J. G., & Huey, R. B. (1998). Evolutionary analyses of morphological and physiological plasticity in thermally variable environments. *American Zoologist, 38*, 545-560.

Komers, P. E. (1996). Behavioural plasticity in variable environments. *Canadian Journal of Zoology, 75*, 161-169.

Kondrashov, A. S., & Houle, D. (1994). Genotype-environment interactions and the estimation of the genomic mutation rate in Drosophila melanogaster. *Proceedings of the Royal Society of London Series B Biological Sciences, 258*, 221-227.

Lande, R. (1995). Mutation and conservation. *Conservation Biology, 9*, 782-791.

Lank, D. B., Smith, C. M., Hanotte, O., Burke, T. A., & Cooke, F. (1995). Genetic polymorphism for althernative mating behavior in lekking male ruff, *Philomachus pugnax. Nature, 378*, 59-62.

Livingstone, F. B. (1992). Polymorphism and differential selection for the sexes. *Human Biology, 64*, 649-657.

Lynch, M. (1994). Neutral models of phenotypic evolution. In L. A. Real (Ed.), *Ecological Genetics* (pp. 86-108). Princeton: Princeton University Press.

Lynch, M., Conery, J., & Burger, R. (1995). Mutation accumulation and the extinction of small populations. *American Naturalist, 146*, 489-518.

Lynch, M., Latta, L., Hicks, J., & Giorgianni, M. (1998). Mutation, selection, and the maintenance of life-history variation in a natural population. *Evolution, 52*, 727-733.

Mackay, T. F. C. (1980). Genetic variance, fitness, and homeostasis in varying environments: an experimental check of the theory. *Evolution, 34*, 1219-1222.

Mackay, T. F. C. (1981). Genetic variation in varying environments. *Genetical Research, 37*, 79-93.

Mackay, T. F. C., & Fry, J. D. (1996). Polygenic mutation in *Drosophila melanogaster*: genetic interactions between selection lines and candidate quantitative trait loci. *Genetics, 144*, 671-688.

Mackay, T. F. C., Fry, J. D., Lyman, R. F., & Nuzhdin, S. V. (1994). Polygenic mutation in *Drosophila melanogaster*: Estimates from response to selection of inbred strains. *Genetics, 136*, 937-951.

Mackay, T. F. C., Lyman, R. F., & Hill, W. G. (1995). Polygenic mutation in *Drosophila melanogaster*: Non-linear divergence among unselected strains. *Genetics, 139*, 849-859.

Mackay, T. F. C., Lyman, R. F., Jackson, M. X., Terzian, C., & Hill, W. G. (1992). Polygenic mutation in *Drosophila melanogaster*: Estimates from divergence among inbred strains. *Evolution, 46*, 300-316.

Martin, N. G., Eaves, L. J., & Eysenck, H. J. (1977). Genetical, environmental and personality factors influencing the age of first sexual intercourse in twins. *Journal of Biosocial Science, 9*, 91-97.

Maynard Smith, J., & Hoekstra, R. F. (1980). Polymorphism in a varied environment: how robust are the models? *Genetical Research, 35*, 45-57.

McGue, M., Vaupel, J. W., Holm, N., & Harvald, B. (1993). Longevity is moderately heritable in a sample of Danish twins born 1870-1880. *Journal of Gerontology, 48*, B237-B244.

Mealey, L., & Segal, N. L. (1993). Heritable and environmental variables affect reproduction-related behaviors, but not ultimate reproductive success. *Personality and Individual Differences, 14*, 783-794.

Medawar, P. B. (1952). *An Unsolved Problem of Biology*. London: H. K. Lewis.

Mousseau, T. A., & Roff, D. A. (1987). Natural selection and the heritability of fitness components. *Heredity, 59*, 181-197.

Mukai, T. (1985). Experimental verification of the neutral theory. In T. Ohta & K.-I. Aoki (Eds.), *Population Genetics and Molecular Evolution* . Berlin: Springer-Verlag.

Mukai, T., Cardellino, R. A., Watanabe, T. K., & Crow., J. F. (1974). The genetic variance for viability and its components in a local population of *Drosophila melanogaster*. *Genetics, 78*, 1195-1208.

Neel, J. V., & Schull, W. J. (1971). Differnetial fertility and human evolution. *Evolutionary Biology, 6*, 363-379.

Parejko, K., & Dodson, S. I. (1991). The evolutionary ecology of an antipredator reaction norm: *Daphnia pulex* and *Chaoborus americanus*. *Evolution, 45*, 1665-1674.

Partridge, L., & Barton, N. H. (1993). Optimality, mutation and the evolution of ageing. *Nature, 362*, 305-311.

Plomin, R., Defries, J. C., McClearn, G. E., & Rutter, M. (1997). *Behavior Genetics*. (3rd ed.). New York: W.H. Freeman.

Provine, W. B. (1971). *The Origins of Theoretical Population Genetics*. Chicago: University of Chicago Press.

Rice, W. R. (1992). Sexually antagonistic genes: Experimental evidence. *Science, 256*, 1436-1439.

Rice, W. R. (1996). Sexually antagonistic male adaptation triggered by arrested female evolution. *Nature, 381*, 232-234.

Riddle, R. A., Dawson, P. S., & Zirkle, D. F. (1986). An experimental test of the relationship between genetic variation and environmental variation in *Tribolium* flour beetles. *Genetics, 113*, 391-404.

Rodgers, J. L., Rowe, D. C., & Buster, M. (1999). Nature, nurture and first sexual intercourse in the USA: Fitting behavioural genetic models to NLSY kinship data. *Journal of Biosocial Science, 31*, 29-41.

Roff, D. A. (1992). *The Evolution of Life Histories. Theory and Analysis*. London: Chapman and Hall.

Rose, M. R., & Charlesworth, B. (1981). Genetics of life history in *Drosophila melanogaster*. I. Sib analysis of adult females. *Genetics, 97*, 173-186.

Rose, M. R. (1982). Antagonistic pleiotropy, dominance, and genetic variation. *Heredity, 48*, 63-78.

Rose, M. R. (1985). Life history evolution with antagonistic pleiotropy and overlapping generations. *Theoretical Population Biology, 28*, 342-358.

Rose, M. R. (1991). *The Evolutionary Biology of Aging*. Oxford, U.K.: Oxford University Press.

Rowe, D. C., & Rodgers, J. L. (1993). Can we identify specific environmental influences on behavioral disorders in children? In Thomas J. Bouchard Jr, E. Peter Propping, & et al. (Eds.), *Twins as a tool of behavioral genetics.* (pp. 205-215): Chichester, England UK.

Scheiner, S. M. (1993). Genetics and evolution of phenotypic plasticity. *Annual Review of Ecology and Systematics, 24*, 35-68.

Schnell, F. W., & Cockerham, C. C. (1992). Multiplicative vs. arbitrary gene action in heterosis. *Genetics, 131*, 461-469.

Schroeder, S. A., Gaughan, D. M., & Swift, M. (1995). Protection against bronchial asthma by CFTR DELTA-F508 mutation: A heterozygote advantage in cystic fibrosis. *Nature Medicine, 1*(7), 703-705.

Segal, N. L., & McDonald, K. B. (1998). Behavioral genetics and evolutionary psychology: Unified perspective on personality research. *Human Biology, 70*(2), 159-184.

Tooby, J., & Cosmides, L. (1990). On the universality of human nature and the uniqueness of the individual: The role of genetics and adaptation. *Journal of Personality, 58*(1), 17-67.

Van Delden, W. (1997). Worldwide latitudinal clines for the alcohol dehydrogenase polymorphism in *Drosophila melanogaster*: What is the unit of selection? In R. Bijlsma & V. Loeschcke (Eds.), *Environmental Stress, Adaptation and Evolution* (Vol. 83, pp. 97-115). Basel: Birkhaeuser Verlag.

van Rhijn, J. G. (1991). *The ruff : individuality in a gregarious wading bird.* London: Poyser.

Waddington, J. L., & Youssef, H. A. (1996). Familial-genetic and reproductive epidemiology of schizophrenia in rural Ireland: age of onset, familial morbid risk and parental fertility. *Acta Psychiatrica Scandinavica, 93*, 62-68.

Williams, G. C. (1957). Pleitropy, natural selection, and the evolution of senescence. *Evolution, 11*, 398-411.

Williams, L. A., & Williams, B. J. (1974). A re-examination of the heritability of fertility in the British peerage. *Social Biology, 21*(3), 225-231.

Wilson, D. S. (1994). Adaptive genetic variation and human evolutionary psychology. *Ethology and Sociobiology, 15*, 219-235.

Winn, A. A. (1998). Is seasonal variation in leaf traits adaptive for the annual plant *Dicerandra linearfolia*? *Journal of Evolutionary Biology, In Press.*

Zirkle, D. F., & Riddle, R. A. (1983). Quantitative genetic response to environmental heterogeneity in *Tribolium confusum*. *Evolution, 37*(3), 637-639.

Chapter 2

GENETIC INFLUENCES ON CHILDBEARING MOTIVATION: FURTHER TESTING A THEORETICAL FRAMEWORK

Warren B. Miller, David J. Pasta, James MacMurray, Donn Muhleman, & David E. Comings

Key words: childbearing, neurotransmitters, social bonding, functional domain, traits, genetic polymorphism

Abstract: In a previous paper (Miller et al., 1999), we developed and tested a theoretical framework linking human childbearing motivation to a neural substrate in the central nervous system. Central to this motivational model was a four-step psychological sequence that we postulated to underlie reproductive decision-making. The sequence begins with motivational traits, which lead to desires, then to intentions, and finally result in instrumental reproductive behaviors. Here, we re-examine and expand upon our framework, generating a new synthesis suitable for testing with genetic and psychological data.

We collected data from an area probability sample of married couples participating in an on-going study of psychosocial aspects of reproductive behavior. Almost 400 male and female respondents provided buccal scrapings for genetic analysis. Based on our theoretical framework, we selected three neurotransmitter-related genetic polymorphisms that we believed were functionally relevant to childbearing motivation, including one for the oxytocin receptor (OXTR), one for the cannabinoid receptor (CNR1), and one for the serotonin transporter (HTT). These genetic variables, together with selected measures of personality, motivation, and conscious desires and intentions, were combined into a causal model. The model was tested using LISREL, with separate covariance matrices for each sex and equality constraints.

The results provide considerable support for our framework. A number of non-hypothesized connections are also observed and our discussion focuses on possible interpretations of these unexpected findings.

1.   INTRODUCTION

In a previous paper (Miller et al., 1999), we developed a theoretical framework that linked the human motivation for childbearing to an underlying neural substrate in the central nervous system (CNS), especially in the areas commonly considered to constitute the motivational and emotional brain. Because of the relative paucity of studies about how the human CNS subserves parental behavior, we drew heavily from relevant research that has been conducted on the domestic rat and extrapolated the findings, using supportive evidence from humans and other animals, to understand the relationship between the CNS and reproductive behavior in humans. We then tested our theoretical framework with data regarding dopamine receptor genes from a sample of over 300 men and women living in northern California.

In the current paper, we continue this work. First we summarize the theoretical framework as previously formulated and expand on certain components of it that are essential to our current purposes. Then we use data from the same sample but related to a different set of neurotransmitters to test the expanded framework. Again our goal is to explicate the connections between the motivational brain and reproduction in humans.

As will become increasingly clear in the next section, our theoretical framework represents an effort to build a bridge across the gap between two distinct, and sometimes seemingly incompatible, levels of explanation: the biological and the psychological. To accomplish this unification, we adopt what is essentially an evolutionary perspective and draw upon the concepts and findings of many different disciplines and subdisciplines. In doing so, we undoubtedly stretch the meaning of some constructs beyond their conventional usage and arrive at some conclusions that are incompletely justified. However, in spite of these potential pitfalls, we are strong believers in the benefits of the "interlocking of causal explanation across disciplines" – what Wilson (1998) has called consilience – and the ultimate strength of theoretical formulations that do successfully build disciplinary bridges.

2.   THEORETICAL FRAMEWORK

Here we summarize our theoretical framework by organizing the discussion around four perspectives. Tinbergen (1951) said there were four ways of answering the question, "Why does an animal behave as it does?" One way of answering is in terms of the evolution of the behavior. A second way is in terms of the behavior's adaptive function. A third way is in terms of the behavior's underlying causation, which to Tinbergen meant primarily its physiology. Finally, the question may be answered in terms of how the behavior develops during the animal's life, beginning with fertilization and extending to old age. We will consider each of these perspectives, and in the course of doing so will incorporate a fifth perspective not emphasized by Tinbergen, namely the psychological. Because we are concerned here with humans and their reproductive behavior, it will increase the explanatory and predictive power of our theoretical framework to include consciousness in the psychological perspective.

2.1   Evolution

A basic assumption of our framework is that reproductive behavior has evolved within each species as an integral part of its overall adaptive strategy. We further assume that because reproduction is so central to the successful survival of each species, the primary components of behavior that enable reproduction to occur with regularity are structurally represented (hard-wired) in the sensory, central processing, and motor areas of the CNS. We believe that this structural representation is as true in human as in other animals in the vertebrate subphylum and that this principle applies at least as well to the motivational and affective components of reproduction as to any other components.

The effect of evolution on brain structure and function has been summarized by MacLean (1990) in his concept of the triune brain. (For an overview of this concept as it applies to emotions and motivations, see Panksepp, 1998.) The triune concept postulates that embedded in the human brain are three layers of brain structures, which represent the legacy of evolutionary progression. The deepest layer consists of the basal ganglia and represents the reptilian legacy. The next layer consists of the limbic system and represents the paleomammalian legacy. The final layer consists of the neocortex and represents the neomammalian legacy. Respectively, these three layers are involved with basic and relatively fixed (instinctive) action tendencies, less fixed but functionally organized emotional and motivational systems, and relatively open, information-processing cognitive systems. The important point for our present purposes is that human reproduction participates in and is an expression of all three of these brain layers. Thus our reproductive behavior, motivations, and cognitions are all constructed from the accumulated pieces of our evolutionary heritage.

2.2   Adaptation

It is the adaptation of organisms to the environment that allows survival. Successful reproduction then assures continuity of the organism's particular type of adaptation (i.e., its adaptive strategy) in the next generation.   As this process goes forward generation after generation, differential survival and differential reproduction shape the structure and behavior of the evolving species and generate its particular design. In short, adaptation produces and defines function.

In our theoretical framework, we make use of the concept of a functional domain. This construct refers to all the structural and behavioral phenomena that serve a single purpose or function. Here we are particularly interested in the reproductive domain. This is a very broad domain that can be subdivided into subdomains, such as sexuality and parenting.   These subdomains can be further subdivided, and this process can be repeated a number of times.   The point is that the functional domain construct is a useful way of grouping and organizing the biological phenomena that serve a common purpose, no matter how molar or molecular these phenomena, and the purposes that they serve, may be.

Since Darwin, there have been many approaches to describing the domains of animal behavior.   The ethologist Scott (1958) described nine categories of behavior.   These included ingestive, shelter-seeking, agonistic, sexual, caregiving, care-eliciting, eliminative, allelomimetic (contagious), and investigative behaviors.   Each of these might be considered a high-level (i.e., relatively abstract) behavioral domain for research purposes. Although these nine categories do appear to cover a very large proportion of common animal behaviors, a problem with this particular taxonomy is that other investigators might reasonably cut up the behavioral pie differently by lumping or splitting categories or by drawing different boundaries.

One approach to identifying a valid taxonomy is to explore the relationship between selected behavioral domains and specific CNS structures and functions. In particular, because emotional/motivational brain systems are so central to the purposes or functions of behavioral domains, anchoring behavior to these systems should help clarify their definition. Further, if we think of these brain systems as constituting a set of emotional/motivational domains, then the theoretical task is to map the two types of domains – emotional/motivational on one hand and behavioral on the other – onto each other, generating a cross-level interlocking of the domain construct.

In this regard, Panksepp (1998) has offered a neurologically-based taxonomy of emotional/motivational processes, each of which is based on specific and to some extent unique CNS circuitry. His categories (and their brief descriptions) include: seeking (the search for food, water, and warmth), lust (the motivation for sex and companionship), nurturance (the motivation to care for offspring), panic (the emotional urge to reunite after separation), fear (the emotional urge to avoid pain and destruction), rage (the emotional urge to deal vigorously with frustration and threat), and play (the emotional urge to interact socially in a way that is vigorous and fun). Panksepp recognizes that additional emotional/motivational categories may be warranted but focuses his discussion on the seven listed. Although a juxtaposition of the nine behavioral domains of Scott with Panksepp's seven emotional/motivational domains reveals important differences and suggests possible omissions on both sides, it is not difficult to recognize the heuristic value of drawing upon both perspectives in the effort to refine and explore the functional domain construct.

In our theoretical framework, described previously (Miller et al., 1999) and elaborated here, we postulated that the two primary adaptive tasks of living organisms, survival and reproduction, and therefore all the functional domains that flow from those tasks, have been addressed in mammals by the development of a broad strategy of social bonding. Although other classes of vertebrates utilize social bonding in various ways, it is most extensively developed in mammals, especially in the primate order, including humans. We further postulated a four-part developmental sequence of bonding that begins after birth with bonding of the infant to the mother (succorant bonding), progresses in childhood with bonding of the child to siblings and playmates (affiliative bonding), continues in adolescence with bonding to a sexual partner or mate (sexual bonding), and finally arrives during late adolescence and early adulthood at bonding of the parent to an infant (nurturant bonding). We believe that this sequence is so crucial to survival and reproductive success that the emotional/motivational dispositions that enable it to unfold in this stepwise fashion across the life course must be represented by extensive neural circuitry within the CNS. In the cited reference, we have illustrated the type of circuitry that may underlie human nurturant bonding by summarizing relevant research findings from studies of the domestic rat.

We also postulated that there are two types of subsystems within the overall neural system that promotes bonding, an affectional subsystem and a security subsystem. Each of these subsystems operates within each of the four main bonding systems. Thus within the succorant domain, feelings of love for the mother reflect the working of the affectional subsystem, while attachment anxiety reflects the working of the security subsystem. Similarly, within the nurturant domain, the corresponding two subsystems generate love for the child and child-care anxiety.

These latter two emotional/motivational states correspond well to maternal factors believed to operate in other animals. Pryce (1992) has described a systems model of the various factors that contribute to maternal motivation in mammals. He identifies four factors that affect the mother's central state of maternal arousability and thereby influence her maternal behavior. These include two factors that promote maternal behavior, maternal attraction to the infant and anxiety about it, and two factors that inhibit maternal behavior, aversion to the infant and novelty-fear of it. Maternal attraction and anxiety closely correspond to the love and anxiety generated within the nurturant domain by the affectional and security subsystems we postulate. On the other hand, aversion (for example, dislike of the infant's clinging behavior) and novelty-fear (generated in inexperienced mothers by the unfamiliar stimulus of an infant) seem to represent conflicts between the nurturant domain and other domains having to do with, for the examples sited, comfort and self-preservation. We will return to this issue of cross-domain conflicts in the next section.

On the basis of our review of the domestic rat research literature, we identified a number of neurotransmitters that seem to perform specific functions within the overall bonding system and its two subsystems. Table 1 gives an overview of how we conceptualize the operation of selected neurotransmitter systems within the nurturant domain. Domain systems are those that act to enhance and coordinate activities within domain-specific circuits. Motive systems act to reward or punish behavior generated by domain-specific circuits. Arousal systems act to alert the organism to environmental factors with domain-specific significance. Finally, executive systems act to activate domain-specific behavior and coordinate activities that have cross-domain implications. In the empirical part of this paper, we will examine genetic aspects of several of the neurotransmitters listed in Table 1 as they relate to human childbearing motivation.

2.3   Causation

Although Tinbergen (1951) considered psychology to be a causal science of behavior, he focused on physiological factors as providing the underlying causal structure of behavior. He did this in part because he was especially interested in innate behavior and in part because he considered the subjective phenomena studied by psychologists to be outside the reach of objective study. Here we take a different position, namely that the underlying causal structure of behavior is both psychological and biological. In fact, we would argue that our understanding of behavior can be greatly strengthened by considering both levels simultaneously, allowing the constructs we use and the questions we ask at each level to contribute to those at the other. Further, we believe that at the psychological level, constructs that embody subjective experiences not only need not be avoided but actually contribute to the cross-level interlocking of understanding.

Table 1. Neurotransmitter systems hypothesized to affect the nurturant domain, organized by type, function, and specific neurotransmitter

Type of Neurotransmitter System	Function	Specific Neurotransitters
Domain	Enhance and coordinate domain-specific CNS processes	oxytocin prolactin vasopressin
Motive:	Motivate behavior	
Hedonic		endorphins other opioids cannabinoids
Fear		glutamate
Arousal:	Alert organism to biologically significant environmental events	dopamine serotonin norepinephrine
Executive:	Coordinate different domain tendencies and activate behavior	γ-aminobutyric acid

2.3.1   Psychology

When trying to explain behavior at the psychological level it is common to use dispositional or trait constructs. Thus we might try to account for the behaviors of a pet cat by saying it is "curious," or "fearful," or "friendly." These are behavioral traits. In humans the use of trait constructs has a long history (Winter, John, Stewart, Klohnen, & Duncan, 1998); these have been used not just in relation to behavior but in relation to affect and cognition as well. Perhaps the most common examples are those of temperament traits in infants and younger children and personality traits in older children and adults, both of which refer to dispositions in broad, complex areas of functioning. Recently, some consensus among psychologists has emerged to the effect that five clusters of personality traits exist in humans (Winter et al., 1998; McCrae & Costa, 1997). Other common examples of the trait construct in humans include general intelligence (I.Q.) and emotional dispositions such as trait anxiety. In general, no matter how broad or narrow and no matter what the referent, the trait construct refers to some way of thinking, feeling, or behaving that is consistent across time and situations.

Miller (1995) has described two broad motivational traits relevant to reproductive behavior, Positive Childbearing Motivation (PCM) and Negative Childbearing Motivation (NCM), which are assumed to be built upon a biologically-based emotional/motivational responsiveness to children. These traits are measured with the Childbearing Questionnaire (CBQ) by asking respondents to indicate the desirability (in the case of PCM) or undesirability (in the case of NCM) of a number of consequences of having a child. The responses are then averaged across the 27 listed items for PCM and across the 20 listed items for NCM. These two broad motivational traits represent the starting point for our psychological-level causative model. Before further explicating this model, however, it will be useful to consider some empirical findings that relate these two motivational traits to our bonding framework and to Pryce's systems model of mammalian maternal motivation.

Although it seems counter-intuitive, we have consistently found PCM and NCM not to be negatively correlated with each other to an appreciable degree. However, a different and instructive picture emerges when the subscales of both main scales are correlated. PCM has five subscales and NCM has four. The full and abbreviated names of these subscales, the number of subscale items, and a representative subscale item (the female version if there is a gender difference) are shown in Table 2. The correlations of PCM and its five subscales with NCM and its four subscales are shown in Table 3. These correlations are based on average scores derived from multiple (in most cases, five) administrations of the CBQ during the course of a longitudinal study of the reproductive behavior of 405 married couples. It is a large subsample of these respondents that provided the genetic data used in our previous paper (Miller et al., 1998) and that will be used for the analyses presented in the empirical section of this paper. The finding in Table 3 which is most interesting and most relevant to our current discussion is that whereas PCM and NCM are uncorrelated with each other, their subscales reveal a different pattern. This is most clearly seen in the two main subscales of NCM, namely Negatives of Child Care (NegChC) and Fears and Worries of Parenthood (FaWPar). Both are strongly and significantly correlated with PCM and all its subscales, but for NegChC, the correlations are <u>negative</u>, whereas for FaWPar they are <u>positive</u>.

We interpret these results to be consistent with Pryce's (1992) discussion of the four factors that contribute to maternal motivation among mammals. Thus the two Pryce factors that promote maternal behavior, attraction to the infant and anxiety about it, correspond to PCM (and all its subscales) and the Fears and Worries of Parenthood subscale of NCM. As they should be, these scales and subscales are all positively correlated. The fact that FaWPar is a subscale of NCM almost certainly reflects the fact that whereas fears and worries promote caretaking behavior, they are perceived (especially in their extreme form) as undesirable. Of the two Pryce factors that inhibit maternal behavior, aversion and novelty-fear, only aversion is represented among the NCM subscales, specifically by Negatives of Child Care. This subscale is made up of items that reflect well the kinds of cross-domain conflicts that

Table 2. PCM and NCM subscale names (abbreviations), item numbers, and representative items (female version).

Scale/Subscale (abbreviation)	Number of Items	Representative Item
PCM		
Joys of Pregnancy, Birth, and Infancy (JoysPBI)	6	Holding and cuddling a baby
Satisfaction of Childrearing (SatChR)	6	Playing with my child
Traditional Parenthood (TradPar)	6	Having a child who will carry on my family traditions
Feeling Needed and Connected (FeelNaC)	5	Living a fuller, more enriched life through my child
Instrumental Values of Children (InstVoC)	4	Knowing that I am fertile
NCM		
Discomforts of Pregnancy and Childbirth (DiscPaC)	2	Experiencing the pain of childbirth
Negatives of Child Care (NegChC)	8	Having to put up with the mess and noise that children make
Fears and Worries of Parenthood (FaWPar)	6	Worrying about the health and safety of my child
Parental Stress (ParStr)	4	Straining our marriage with a baby

human parents feel, as illustrated by not wanting to put up with the noise and mess of children. The other two subscales appear to represent factors that are peculiarly human: anticipation of discomfort during pregnancy and birth and concern for the effect of childbearing on the marriage. These subscales are correlationally neutral with respect to PCM, but probably also reflect cross-domain conflict. Novelty-fear of newborn probably has an inexact correspondence to the lack of confidence first-time parents feel about their child and as such is most likely subsumed within the FaWPar subscale.

Table 3.   Correlations of PCM and its five subscales with NCM
and its four subscales.

Positive Childbearing Motivations[a]	Negative Childbearing Motivations[a]				
	NCM	DiscPaC	NegsChC	FaWPar	ParStr
PCM	-.028	.003	-.254**	.223**	.031
JoysPBI	-.172**	-.121**	-.401**	.119**	-.039
SatChR	-.052	.020	-.302**	.209**	.036
TradPar	.015	.034	-.126**	.172**	.023
FeelNaC	.057	.036	-.099**	.242**	.038
InstVoC	.065	.070*	-.124**	.228**	.098**

* = p < .05, ** = p < .01

[a]These scores are averaged across 5 (n = 628), 4 (n = 72), 3 (n = 74), 2 (n = 17),
or 1 (n = 19) measurements.

By way of summary, if we relate these trait findings back to our
discussion of the affectional and security subsystems within the nurturant
bonding domain, it appears that PCM and its subscales all reflect the
affectional subsystem. In fact, a good argument can be made that the first
four subscales represent how the affectional subsystem operates across the
parental life-course, with JoysPBI, SatChR, TradPar, and FeelNaC reflecting
the different mental representations of the affectional subsystem within the
nurturant domain during early, early-middle, late-middle, and late parenting,
respectively. On the other hand, it appears that NCM is a more complex
measure, reflecting the security sub-system in one subscale (FaWPar) and
cross-domain conflicts in at least one, and very probably all three, of the
other NCM subscales. This complexity of NCM probably reflects the
derivative nature of these two broad motivational traits, i.e., that they
derived during development from more prototypical traits. We will consider
this issue further in the next section.

We return at this point to the explication of our psychological-level
causative model. In studying human behavior, we are fortunate not to be
limited to the use of traits. We can introspect, and we can ask our

respondents to introspect, about mental phenomena that are antecedent to behavior. Applying this sort of approach to reproductive behavior, Miller (1994) has developed a four-step psychological model, which we will refer to as the TDIB model. The T represents the motivational traits that we have just discussed. The D and I represent various types of reproductive desires and intentions, where desires represent what the individual consciously wants (e.g., having a child, having a certain number of children, and having a child at a certain time) and intentions represent those desires as constrained by reality, especially by what the individual's partner wants. Finally, the B represents reproductive behavior. In his research, Miller has shown that intentions are the best predictors of proceptive behavior (Miller & Pasta, 1995a), which is behavior designed to achieve conception (Miller, 1986); that desires are the best predictors of intentions (Miller, 1994; Miller & Pasta, 1994); and that motivations are the best predictors of desires (Miller & Pasta, 1993; 1994).

The TDIB model we have just described needs to be expanded in order to represent more faithfully the actual antecedents to human reproductive behavior. Just as there are different types of antecedent motivation, so there are different types of desires and intentions. We present the overall model schematically in Figure 1, which shows the relationship of childbearing motivations to three types of desires, their corresponding intentions, and the resultant behavior. All the components of this model have been examined extensively in the references cited in the previous paragraph. We will use a somewhat restricted version of this model to test our genetic data in the empirical section of this paper.

2.3.1   Biology

In our previous paper (Miller et al., 1999), we discussed in some detail the neural circuitry that appears to underlie nurturant domain-specific motivational processing in the domestic rat. The CNS area involved in this circuitry is the ventral forebrain and includes the following five structures: the medial preoptic area, the ventral bed of the stria terminalis, the ventromedial nucleus of the hypothalamus, the ventral tegmental area, and the amygdala. We also discussed how specific neurotransmitter systems located within these areas participate in rat nurturant behavior. These systems include the domain, motive, and arousal systems described above and summarized in Table 1. In humans, there is comparable, on-going work

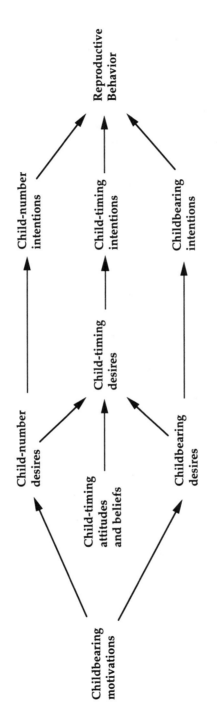

Figure 1. A schematic representation of the pathways through which childbearing motivations, three types of fertility desires, and three corresponding types of fertility intentions affect reproductive behavior.

on the biological substrates of motivation (Duffy, 1997; Kalivas & Barnes, 1993), focusing on the same or corresponding CNS areas.

We hypothesize that this ventral forebrain circuitry, together with projections to the prefrontal cortex, the hippocampus, cortical areas adjacent to the motivational processing area, the hypothalamus, the periaqueductal grey area, and the brain stem, are all involved in the formation, control, and expression of domain-specific motivational traits such as PCM and NCM. To some extent traits such as emotional/motivational responsiveness to babies are hard-wired but it takes experience, reward and punishment, and finally memory formation to build those traits into ones that are smoothly integrated into the regulation of adult human behavior.

As we have suggested by the TDIB model discussed above, the translation of those traits into behavior in humans also involves several steps that require consciousness. The neural substrates underlying consciousness are only just beginning to be understood. In our previous paper, we approached this subject by expanding upon a perception/central-processing/response model of consciousness that drew heavily upon LeDoux's (1996) treatment of working memory, emotions, and their neural substrate. Here we simply note that the motivation-desire-intention sequence may be thought of as involving progressively greater degrees of cortical regulation and progressively lesser degrees of ventral subcortical processing.

A distinction that is helpful at this point is that between implicit and explicit motivation (for a discussion, see Winter et al., 1998). For a number of reasons, both neurological and psychological, human motivations may be inaccessible or only partially accessible to consciousness. PCM and NCM are measures of explicit motivations because they require conscious verbal responses. It is very likely that there are other, more implicit childbearing motivations, i.e., those that are less accessible to consciousness. Indeed, elsewhere in this volume, Miller and Pasta (1999) consider possible indicators of such motivations. This distinction between explicit and implicit motivation may also be seen as involving more or less cortical involvement and less or more ventral subcortical processing.

An additional hypothesis we proposed in our previous paper is that there is some common CNS circuitry underlying the emotional/ motivational aspects of both nurturant bonding and the three other types of bonding we have described. We also propose that there is some common circuitry underlying both the affectional and the security components of these four bonding systems. These ideas are consistent with MacDonald's (1991) proposal that human goal-directed behavior is in the service of evolved motive dispositions (i.e., motivational traits) that are arranged in the CNS in a hierarchy characterized by (among other things) more or less domain specificity. A number of evolutionary-oriented researchers (Gazzaniga, 1985; Tooby & Cosmides, 1992; Buss, 1995) have argued that the human brain has a modular structure and that the human mind is based on a series of domain-specific adaptive mechanisms. Here we have suggested what both the psychological and biological components of such mechanisms might be with respect to bonding in general and nurturance more specifically. In short, we have described a motivationally-centered, nested set of bio-behavioral modules that function to promote bonding.

2.4   Development

All of the perspectives we have discussed to this point have important developmental aspects. It is a biological truism to say that an organism's evolutionary heritage is reflected in its development and so it should be kept in mind that the developmental processes we consider have been built up and constrained by evolution. With respect to adaptation, we have already described how the bonding systems that we postulate have been developmentally keyed, that is to say are designed to address bonding requirements as the organism progresses from being a baby to parenting. Here we will focus primarily on one feature of the causal processes we have described: how the emotional/motivational traits that are central to the bonding mechanisms (modules) develop from infancy to adulthood.

We have already mentioned temperament and personality as examples of the trait construct. In this regard, Rothbart, Derryberry, and Posner (1994) have proposed a developmental model that links newborn, older infant, and younger child temperament to older child and adult personality. Their approach provides a good starting point for understanding how traits change and diversify during the life course and how adult traits may be interpreted as derivative of child and even infant traits. Although their model is more complex than what we now present and certain infant temperament traits such as Activity Level and Rhythmicity that appear unrelated to emotional/motivational factors are not included, we use it as an example of how emotional/motivational traits might develop during the lifecourse. In their approach they suggest that three temperament factors commonly found in infants are plausible antecedents to three of the big five personality clusters found in adults. Specifically, Surgency (also called Approach or Positive Affect) can be related to the adult personality trait cluster of Extraversion, Negative Affectivity can be related to the adult trait cluster of Neuroticism (also called Emotional Stability), and Effortful Control can be related to the adult trait cluster of Conscientiousness (also called Dependability). They also suggest that a late developing temperament trait which they call Positive Social Orientation or Affiliation may be the antecedent to a fourth adult trait cluster, Agreeableness. (The fifth adult trait cluster of Openness is not described as having an antecedent in temperament.) There are obviously other ways to organize the array of results in these two related fields, but the important point is that a substantial degree of continuity between child and adult traits probably exists on the basis of developing neural substrates. To be sure, as Rothbart, Derryberry, and Posner discuss, these traits are shaped by events and experiences during development. But they are also undoubtedly shaped by constraints imposed by the underlying neural substrate.

Related to this is the relationship between adult personality traits and childbearing motivation. In two previous studies, Miller (1981, 1992) used selected personality traits measured by the Personality Research Form (Jackson, 1984) to predict PCM and NCM. In the first study he found that Nurturance predicted both PCM and NCM and Succorance predicted PCM. In the second he again found that Nurturance predicted both PCM and NCM and that both Affiliation and Autonomy (negatively) predicted PCM. Only Nurturance was tested in both studies, although Succorance and Autonomy (with a -0.68 correlation in Jackson's research) are very closely related. These findings support the proposition that the neural substrate underlying at least three of the four bonding systems described above – namely the succorant, affiliative, and nurturant systems – contribute to both adult personality traits and childbearing motivations.

We propose, therefore, that there is a developmental sequence in humans that relates to the domain of bonding. At the psychological level this sequence would begin with very broad temperament traits such as Approach and Negative Affectivity differentiating across time into clusters of personality traits such as Extraversion and Emotional Stability. Perhaps then, as suggested by Rothbart, Derryberry, and Posner, there would be some late developing temperament trait such as Positive Social Orientation differentiating into adult Agreeableness. Finally, the sequence would conclude with more specific traits from within the adult trait clusters of Extraversion, Agreeableness, and Neuroticism differentiating into specific motivational traits such as PCM and NCM.

Exactly how this last development would occur needs to be further investigated. We know already (Costa & McCrae, 1988) that two of Jackson's bonding-related measures, Affiliation and Nurturance, are strongly correlated with both Extraversion and Agreeableness adult trait clusters, although Affiliation more with Extraversion and Nurturance more with Agreeableness. We also know from Costa and McCrae that Jackson's measure of Succorance is correlated with Neuroticism. However, in our culture, where the childlike bonding style reflected in Succorance is not acceptable, that finding is not surprising. If one uses a six rather than a five factor model of personality (Jackson, Paunonen, Fraboni, & Goffin, 1996), the trait cluster of Neuroticism is replaced by one called Independence, which includes both Autonomy and (negatively) Succorance. If we assume from Miller's earlier work (1981, 1992) that Affiliation, Nurturance, and probably Succorance all contribute to the affectional component of PCM and that some unmeasured fear-related traits such as Harmavoidance or Trait Anxiety contribute to the security component of NCM, then a reasonable developmental sequence can be identified. Broad temperament traits of infancy differentiate into clusters of adult personality traits. Within these clusters, specific bonding-related traits differentiate into motivational traits related to childbearing. In our view, these developmental continuities reflect the influence across time of a common neural substrate. Of course, there are discontinuities as well. Each bonding-related trait probably has its own unique substrate and, as we have indicated, non-bonding domains make important contributions, especially in the case of NCM.

We have said nothing about the developmental structure of childbearing desires and intentions, in large part because virtually no systematic research has been conducted on this topic earlier than late adolescence. Although it can be argued that desires and intentions do not become well formed before that period, a 1993 pilot study (Miller, unpublished results) conducted with 18 male and female youth (ages 10 to 14) indicates otherwise. Many of these youth had specific desires and intentions, suggesting that it might well be instructive to study changes in these conscious elements and their associated antecedent traits beginning with, or even before, puberty.

2.4   Synthesis

We both integrate and summarize the discussion of our theoretical framework in Figure 2, which represents the proposed causal sequence leading to behavior within the nurturant domain. The right side of the figure – including the columns for motivational traits and for conscious desires/intentions – is modified from Figure 1, showing both PCM and NCM but including only those desires and intentions most directly influenced by them and excluding behavior completely.   Antecedent to the two motivational traits are two categories of personality traits, each hypothesized to affect both of them. We assume that affectional traits are antecedent primarily to PCM and security traits are antecedent primarily to NCM, but we know from earlier work that crossover takes place (e.g., Nurturance predicts NCM [Miller, 1992] and Harmavoidance predicts PCM [Miller, 1981]). Finally, the figure indicates that three categories of neurotransmitter systems are all hypothesized to affect both affectional and security traits.

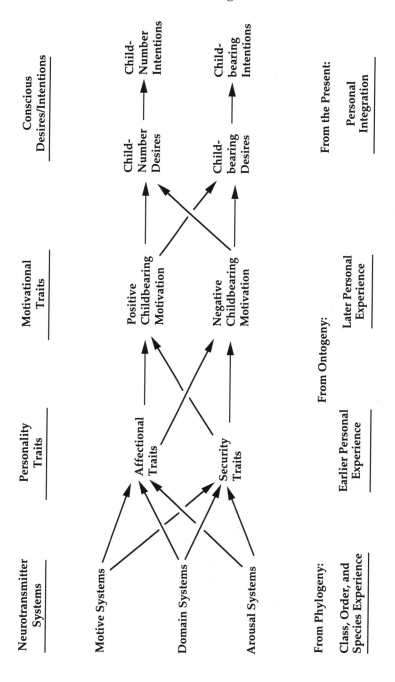

Figure 2. A synthesis of the theoretical framework as applied to the nurturant domain.

Figure 2 presents a model that is cross-sectional in nature, that is to say, one that can be tested with data collected at one point in time. However, there is a longitudinal structure that is implicit in the model as well, one that is described in the captions across the bottom. The neurotransmitter systems represent accumulated phylogenetic experience, the traits represent effects on the neural substrate of earlier and later personal experience, and conscious desires and intentions reflect the integration of motivations within and across domains in the context of the present situation. This longitudinal structure embodies both the evolutionary and the developmental perspectives that we have discussed.

3.   METHODOLOGY

Beginning in November 1988, a modified area probability sample of 401 married couples living in Santa Clara County, California was generated, using a reverse telephone directory and telephone recruitment (Miller, 1992). By design, 203 couples had no children and 202 couples had one child. Mean age of the wives was 29.6 and of the husbands was 31.6. The obtained sample was predominantly middle class: mean years of education and mean income (in $1000s) was 15.9 and 25.7 for the wives and 16.7 and 45.9 for the husbands. Eighty-one percent of the sample was non-Hispanic, European-American (i.e. white) and the remainder African-American, Asian-American, or Hispanic-American. Parity at the time of the final follow-up five years later for those couples remaining in the study was as follows: 78 (22%) had no children, 102 (29%) had one child, 151 (43%) had two children, and 22 (6%) had three or more children.

Each respondent was initially interviewed in person and then administered a battery of self-report inventories. These included the Childbearing Questionnaire (Miller, 1995) and two scales from Jackson's Personality Research Form (1984): Nurturance and Affiliation. Four follow-up interviews were conducted, one, two, three and one half, and five years after the initial interview, at which times the Childbearing Questionnaire was readministered. At each of the five interviews, respondents were also asked a single question about the number of children they desired (Miller and Pasta, 1993), two questions about the strength and certainty of their childbearing desires, and two similar questions about the strength and certainty of their childbearing intentions (Miller, 1994). Data for child-number intentions were collected only during the last two interviews and were judged insufficient for use in the analyses reported below.

We created five psychological variables derived from our formulation of the motivational antecedents to reproductive behavior – see Figures 1 and 2. These variables were Positive and Negative Childbearing Motivation (PCM and NCM), Child-Number and Childbearing Desires, and Childbearing Intentions. In order to reduce error variance and maximize variance due to possible genetic influence, we averaged the five measurements of each of these five motivational antecedent variables for each respondent.

We obtained a buccal scraping sample for purposes of genetic analysis at the last follow-up interview. In order to avoid confounding any findings due to an association between genotype and race, we included only non-Hispanic, European-American (white) respondents. Among those eligible for inclusion and not lost to follow-up, 536 (84.5%) agreed to participate. Of those, 430 (80.2% of 536) actually provided a tissue sample.

We selected one specific neurotransmitter from Table 1 to represent each type of neurotransmitter system included in Figure 2. Only neurotransmitters with functionally relevant polymorphisms of the transmitter itself, its receptor, or its transporter were considered, the rationale being that individual variation based on such a polymorphism would possibly affect the next step in the sequence shown in Figure 2. To represent the domain systems, we selected the oxytocin receptor (OXTR) polymorphism because oxytocin has been shown to participate in maternally relevant neural systems in voles and rats (Insel, 1997). Simmons, Clancy, Quan, and Knoll (1995) have localized and characterized OXTR and Michelini, Urbanek, Dean, and Goldman (1995) have described a dinucleotide repeat with two primary alleles at 276 and 278 base pairs. In our experience (Comings & MacMurray, unpublished data), the alleles varied from 272 to 280 base pairs and we observed a negative linear relationship between the length of the alleles and the personality trait of Extraversion. Because of the possible relationship between this phenotype and Nurturance, as discussed above, we planned a linear genotype coding based on the number of base pairs in each allele.

To represent the motive systems, we selected the cannabinoid receptor (CNR1) polymorphism. Arachidonyl-ethanolamide has been identified as the natural ligand of this system (Devane et al., 1992) and delta-9-tetrahydrocannabinol (THC), a naturally occurring compound that stimulates this system, has been shown to affect sexual motivation and emotionality (Navarro, Fernández-Ruiz, de Miguel, et al., 1993). The brain cannabinoid receptor has been identified and cloned by Matsuda, Lolait, Brownstein, Young, and Bonner (1990) and a trinucleotide repeat polymorphism has been described by Dawson (1995). Based on our prior experience (Comings, Muhlman, Gade, et al., 1997), genotypes where both alleles have five or more repeats are associated with different hedonic behaviors compared with the other genotypes. Therefore, we planned to code according to whether the respondent's genotype was ≥ 5 for both alleles ($\geq 5/\geq 5$) or not.

To represent the arousal systems, we selected the serotonin transporter (HTT) polymorphism. There is now considerable evidence that the serotonin system modulates the fear response in animals, including humans (Charney, Woods, Krystal, & Heninger, 1990; DeVry, Schreiber, Glaser, & Traber, 1992). The promoter of HTT is regulated by an interplay between positive and negative regulatory elements (Lesch, Balling, Gross, et al., 1994). A unique GC-rich repetitive sequence is located in the proximal 5'-regulatory region of the HTT gene, which displays a tetrastrand-like structure. The latter silences transcriptional activity in neuroserotonergic cells and contains positive response elements (Heils, Teufel, Petri, et al., 1996). These authors also reported a common deletion polymorphism of this repetitive element. Expression studies in human cell lines have shown that the long (non-deletion) allele was associated with a far greater rate of expression than the short allele. Further, Lesch, Bengel, Heils, et al., (1996) showed that the short allele was associated with anxiety-related traits in humans. Based on this literature, we planned to code both for the presence or not of the short allele (11 and 12 versus 22, where 1 = the short allele) and for high dominance (11 versus 12 and 22).

Data analysis proceeded as follows. We first examined the genotype frequencies in our sample. Because the tissue produced by the buccal scraping technique is less compatible with the polymerase chain reaction and other laboratory procedures than other tissues (e.g., white blood cells), genotyping was successful with a smaller proportion of our sample for some genes compared with others. We were able to genotype 181 respondents (males and females combined) for OXTR, 235 respondents for CNR1, and 188 respondents for HTT.

We next examined the correlations between the coded gene data and both the personality and motivational traits included in our model – see Figure 2. Nurturance was our first affectional trait because the model deals with the nurturant domain. However, because we wanted to test the developmental aspect of our overall framework – especially the part which stipulates that the nurturant system is built upon both the neural substrate and life experiences of earlier appearing bonding systems – and because we had a good measure of Affiliation in our data, we added that as a second affectional trait. Unfortunately, our data set had no measures, such as Harmavoidance or Trait Anxiety, that could be considered to represent the security traits in our model. Thus the correlations we examined were between the three gene variables, on one hand, and Nurturance, Affiliation, PCM, and NCM, on the other.

The final step in the analysis was to test the model shown in Figure 2. We used LISREL (Jöreskog & Sörbom, 1996) and the method of maximum likelihood, utilizing this program's linear structural equation capacity but without creating any latent variables. Because of the different Ns for the genotype variables, the covariance matrix was calculated with pair-wise missing data. The Ns of the motivational trait and desires/intentions variables were all 810, this being the number of husbands and wives in the original study. As explained previously, these motivational trait and desires/intentions scores were averaged across the entire study. Due to varying drop out points, these scores were averaged across 5 (N=628), 4 (N=72), 3 (N=74), 2 (N=17), or 1 (N=19) measurements.

In order to estimate the model shown in Figure 2, we began with specific hypothesized connections. We let CNR1 predict both Affiliation and Nurturance on the assumption that the cannabinoid system played a role in the reward associated with both affiliative and nurturant bonding. We let OXTR predict only Nurturance because the oxytocin system appears to be associated primarily with the reward aspects of sexuality and parenting (Insel, 1997). We let HTT predict NCM directly, given that our data did not contain an intermediary personality trait such as Harmavoidance or Trait Anxiety. Based on earlier work (Miller, 1992), we let Nurturance predict both PCM and NCM but we let Affiliation predict only PCM. We let both PCM and NCM predict both types of desires. Finally, we let Childbearing Desires predict Childbearing Intentions. Hypothesized connections that were not significant (t<1.97, p>.05) were dropped and non-hypothesized connections (as suggested by modification indices) that were significant (t≥1.97, p≤.05 if the sign was as expected, t≥2.60, p≤.01 if the sign was not as expected) were added, and the entire model was re-estimated.

4. RESULTS

Tables 4, 5, and 6 show the genotype and allelic frequencies for the OXTR, CNR1, and HTT polymorphisms. For OXTR and CNR1, both of which were hypothesized to predict Nurturance, Tables 4 and 5 also show the mean scores and standard deviations on that trait variable. In Table 4, these scores are shown for each allelic combination, arranged ordinally as coded (the single 1/4 allelic combination could not be reliably placed in the

Table 4. Genotype Frequencies, Mean Nurturance Score and
          Standard Deviation, and Variable Coding for Oxytocin
          Receptor (OXTR) Polymorphism.

Number of Dinucleotide Repeats, Allele 1/Allele 2	Frequency (%)		Mean Nurturance[a]	S. D. Nurturance[a]	Variable Coding
1 / 1	2	(1.1)	1.00	1.41	1
1 / 4	1	(0.6)	2.00	--	b
2 / 2	12	(6.6)	3.88	3.34	2
2 / 3	2	(1.1)	0.00	1.41	3
3 / 3	99	(54.7)	1.77[c]	2.92	4
3 / 4	31	(17.1)	1.97	2.48	5
4 / 4	26	(14.4)	1.73	2.78	6
4 / 5	1	(0.6)	1.00	--	7
5 / 5	7	(3.9)	.00	2.71	8
	181	(100.1)	1.84	2.86	

[a]ANOVA F ratio = 1.299, p = 0.247
[b]This genotype was not coded
[c]One respondent of this genotype did not complete the measure of Nurturance

coding sequence and so was omitted from the correlational analyses that follow). In Table 5, because the number of allelic combinations was so great, the frequency data are presented by allelotype and the frequency data and Nurturance scores for the genotypes as coded ($\geq 5/\geq 5$ versus not $\geq 5/\geq 5$) are given in the footnote. For HTT, which was hypothesized to predict NCM, Table 6 also shows the mean scores and standard deviations on that trait variable for each of the three genotypes. Because respondents homozygous for the deletion allele (i.e., the 1/1 genotype) showed higher NCM, we coded the HTT variable as high dominant (1/1 versus not 1/1) in the analyses that follow.

Table 5. Allelic frequencies[a] for canabinoid receptor (CNR1) polymorphism.

Number of Trinucleotide Repeats, Alleles 1 and 2	Frequency (%)	
1	9	(1.9)
2	2	(0.4)
3	5	(1.1)
4	139	(29.6)
5	25	(5.3)
6	60	(12.8)
7	88	(18.7)
8	138	(29.4)
9	3	(0.6)
10	1	(0.2)
	470	(100.0)

[a]Genotype frequencies: <5/<5 and <5/≥5 combined = 124 (52.8%) and ≥5/≥5 = 111 (47.2%). Mean Nurturance (S.D.) for <5/<5 and <5/≥5 combined = 2.18 (2.65) and for ≥5/≥5 = 1.33 (2.90).

Table 6. Genotype frequencies and mean NCM score and standard deviation for serotonin transporter (HTT) polymorphism.

Allele 1/Allele 2	Frequency (%)		Mean NCM[a]	S.D. NCM[a]
1/1	39	(20.7)	53.4	8.03
1/2	77	(41.0)	50.1	8.30
2/2	72	(38.3)	51.6	7.50
	188	(100.0)	51.3	8.00

[a]ANOVA F ratio = 2.245, p value = 0.109

Tables 7, 8, and 9 show correlation coefficients overall and by sex for the coded versions of OXTR, CNR1, and HTT with the four trait variables included in the estimated model. The genotypes of OXTR with more dinucleotide repeats are associated with lower Nurturance in the overall sample. The male and female subsamples are too small to show a significant correlation and are essentially equivalent in their association with Nurturance. The genotypes of CNR1 with five or more trinucleotide repeats on both alleles are associated with lower Nurturance and Affiliation in the overall sample. Interestingly, in the two gender subsamples that association is significant for Nurturance only in females and for Affiliation only in males. The 1/1 genotype of HTT is associated with low NCM but with a p value between .10 and .05. However, in the female subsample, the p value of this association drops just under .05. There is also a significant association between the 1/1 genotype and low affiliation for females. Additional analyses of the 1/1 genotype with the subscales of NCM revealed that only the subscale that reflects the security subsystem – Fears and Worries of Parenthood – was significantly correlated. The correlation coefficients were .161 (p=.028) overall, .107 (p=.290) for males, and .228 (p=.032) for females.

Table 7. Correlations (p values) between OXTR 1-8 and PCM, NCM, Nurturance and Affiliation, overall and by sex.

Traits	OXTR					
	Overall (N = 179)		Males (N = 88)		Females (N = 91)	
PCM	.005	(.947)	-.031	(.775)	.058	(.585)
NCM	-.037	(.624)	-.132	(.219)	.034	(.745)
Nurturance	-.148	(.049)	-.131	(.225)	-.126	(.233)
Affiliation	.012	(.879)	-.012	(.914)	.070	(.511)

Figure 3 shows a schematic representation of the connections in the estimated LISREL model, along with the standardized estimated parameters (S.E.P.) and associated t-values of these connections. Many of the connections are as hypothesized, these being shown by solid lines in the figure. There are five non-hypothesized connections, shown in the figure by

Table 8. Correlations (p values) between CNR1 ≥5/≥5 and PCM, NCM, Nurturance, and Affiliation, overall and by sex.

Traits	CNR1					
	Overall (N =235)		Males (N =115)		Females (N =120)	
PCM	-.103	(.116)	-.153	(.103)	-.057	(.533)
NCM	.015	(.817)	-.046	(.628)	.076	(.409)
Nurturance	-.152	(.020)	-.078	(.411)	-.220	(.016)
Affiliation	-.137	(.036)	-.223	(.017)	-.135	(.142)

Table 9. Correlations (p values) between HTT 1/1 and PCM, NCM, Nurturance, and Affiliation, overall and by sex.

Traits	HTT					
	Overall (N = 188)		Males (N = 99)		Females (N = 89)	
PCM	-.047	(.519)	-.067	(.510)	-.012	(.912)
NCM	-.131	(.074)	-.107	(.294)	-.209	(.049)
Nurturance	.054	(.463)	.093	(.365)	.114	(.286)
Affiliation	-.120	(.101)	.021	(.841)	-.252	(.017)

dashed lines. First, Child-Number Desires predicts Childbearing Desires. This is not surprising and, in fact, we have hypothesized a bidirectional effect between these two variables in earlier work (Miller & Pasta, 1995b). Here we elected to allow the modification indices to indicate whether one direction or another (or even both) fitted best for this bivariate relationship. The obtained result might well have been different if we had a Child-Number Intentions variable being predicted by Child-Number Desires. Second, Affiliation predicts NCM. Related to this is the one hypothesized connection that does not occur: Nurturance predicting NCM. It appears that one of these connections replaces the other, which is not especially surprising in a multivariate context given that Affiliation and Nurturance have generally similar bivariate correlations with NCM (-.267 and -.179, respectively). The other three non-hypothesized connections are CNR1 ≥5/≥5 predicting low Child-Number Desires, HTT 1/1 predicting low Affiliation, and Nurturance predicting high Childbearing Desires.

Two other connections were allowed in the psi matrix to accommodate non-causal correlations between variables. These are Childbearing Desires correlating with Childbearing Intentions (S.E.P. = .07, t-value = 3.10) and NCM correlating with Nurturance (S.E.P. = -.18, t-value = -5.81). We interpret the first of these to represent a method correlation, i.e., one based on the similarity of the two questions and the response categories that were used to measure these constructs. The second correlation suggests that there is an overlap in the meaning of these two constructs.

The goodness of fit of the overall model is high. With 23 degrees of freedom, the chi-square is 26.12 (p=.30). The root mean square error of approximation (RMSEA) is .012, with a p-value for a test of close fit (RMSEA <.05) = 1.

5.   DISCUSSION

These results are generally confirmatory of the proposed theoretical framework. The motivational trait/desires/intentions part of the model performed as anticipated. The relationship between the two personality traits and the two motivational traits was close to expectations except that it was Affiliation and not Nurturance that predicted NCM. We interpret this difference from our previous findings (Miller, 1992) as resulting from the partialling effect of the Nurturance/NCM psi matrix correlation, a relationship that was not possible in the regression analysis conducted previously. Finally, the relationships between the neurotransmitter systems and the personality/motivational traits were as hypothesized: the oxytocin system was specific to Nurturance, the cannabinoid system affected both Affiliation and Nurturance, and the serotonin system, presumably acting through some unmeasured fear-related personality trait, affected Negative Childbearing Motivation. Separate correlations indicate the serotonin system especially affected the most fear-based subscale of NCM, Fears and Worries of Parenthood.

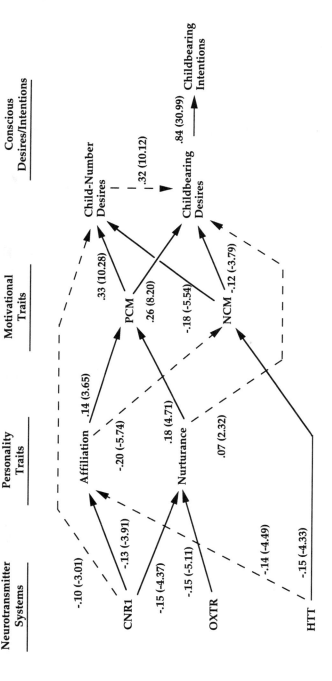

Figure 3. A schematic representation of the estimated model. Numbers are standardized parameter estimates (t-values, with 1.97 = p<.05 and 2.60 = p<.01). Dashed lines indicate non-hypothesized connections.

Each of the three non-hypothesized connections that appear in the left hand portion of Figure 3 informs about potential deficiencies in the theoretical framework. The connection between CNR1 and Child-Number Desires suggests that the cannabinoid system may operate through some unidentified trait(s) to affect the number of children desired. A strong candidate for one such trait would be gender role orientation. There is a body of research supporting the notion that human feelings and cognitions regarding gender roles are rooted in our hominoid ancestry (Clark, 1998), making it feasible that specific neurotransmitter systems, as well as other features of CNS organization, are associated with gender role orientation. A study by Udry, Morris, and Kovenock (1995) indicated that a combination of prenatal and adult androgen exposure in a sample of women affected their adult gendered behavior, where the latter included several measures of gender role orientation. We also know that the number of children desired by married individuals is affected by their gender role orientation (Miller, Turchi, & Pasta, 2000). These lines of evidence support the idea that gender role related traits have a biological base and are good candidates for explaining the CNR1/Child-Number Desires connection.

The connection between HTT and Affiliation is somewhat surprising, in part because we selected the serotonin system for its presumed effect on the modulation of fear, and in part because the sign on the coefficient is negative (both Affiliation and HTT are negatively related to NCM, so we would expect HTT to be positively related to Affiliation). We interpret the unexpected connection to mean that there is some fear component to the affiliative disposition, at least as measured by the Personality Research Form scale. Looking over the items that comprise that scale, especially the negatively scored ones, it is not too difficult to imagine that some of these items tap into people's fear of being apart from the group or alone. Indeed, McClelland, Koestner, and Weinberger (1989) have stated that need Affiliation contains fear of rejection as a major component. Thus the HTT/Affiliation connection reminds us that, as shown in Figure 2, the arousal neurotransmitter systems need not be related just to affectional or security traits: they can be related to both. This was certainly our finding with the variables from the dopamine system, another neurotransmitter involved in arousal, that we studied previously (Miller et al., 1999).

The unexpected connection between Nurturance and Childbearing Desires affirms that there are pathways between these two variables other than through the explicit motivational traits. A strong candidate in this case would be some implicit motivational trait. Elsewhere in this volume (Miller & Pasta, 2000), we argue that a group of developmental variables (age at first sexual intercourse, age at first pregnancy, number of prior pregnancies, and – in women – age at menarche) be considered as measuring an implicit motivational variable and show that it has a substantial effect on the strength of wanting not to get pregnant (a type of childbearing desires variable). Here we suggest that a variable of this sort might very well be acting between Nurturance and Childbearing Desires. In fact, because the aspect of nurturance that we measure here is itself explicit, if we had an implicit measure similar to the one just described in our present data, we would expect OXTR to predict it. It may be that what we are approaching with these ideas is the area of sexual bonding, a part of the overall bonding framework that as yet is not well integrated into our current model.

Although the sample in this study is too small to attempt the construction of separate male and female LISREL models, there are sex differences in two sets of bivariate findings that deserve comment. First, we observed that the cannabinoid system variable was a stronger correlate of Affiliation in men and of Nurturance in women. This parallels an earlier study (Miller, 1992) in which PCM was found to be predicted by Affiliation in men and by Nurturance in women. The two sets of findings together suggest that males and females follow different developmental pathways in the acquisition of positive childbearing motivation and that part of that difference involves the cannabinoid system. Second, we observed that HTT was a stronger correlate of both NCM (and especially its fear-based subscale) and Affiliation in women than it was in men. We have commented previously (Miller et al., 1999) on the importance of the fear motive system, working in conjunction with the hedonic motive system, in the development of concern for others and, ultimately, altruism. We wonder, therefore, if the apparent sex difference in the HTT-to-NCM connection also relates to sex differences in altruism, especially as these reflect relationships with and the care of children.

6. CONCLUSION

The common parlance expression "maternal instinct" attests to the popularity of the notion that parental feelings and behavior have important roots in biology. Here, and in our previous, companion paper (Miller et al., 1999), we have proposed a theoretical framework for understanding the links between the biological and psychological levels in the determination of childbearing motivation and behavior. Our results have tended to confirm some of the specifics of our bonding framework, such as the relevance of certain neurotransmitters that we hypothesize to participate in domain, motive, and arousal systems. They have also lent support to the idea that the nurturant domain is built upon the neural substrate and/or experiences of the affiliative domain, and thus to the idea of a developmental sequence within the bonding domains. Finally, our results have also raised important issues, such as how the neural substrate to bonding affects related psychological areas like role preferences, how the distinction between implicit and explicit motivation fits into our overall framework, and how nurturant traits emerge and develop during infancy and childhood. Perhaps as important as our specific findings has been our demonstration that the construction and testing of a consilient framework linking biology and psychology is both feasible and instructive.

REFERENCES

Buss, D.M. (1995). Evolutionary psychology: A new paradigm for psychological science. *Psychological Inquiry, 6*, 1-30.
Charney, D.S., Woods, S.W., Krystal, J.H., & Heninger, G.R. (1990). Serotonin function and human anxiety disorders. *Annals of the New York Academy of Sciences, 600*, 558-573.
Clark, G.A. (1998). Human monogamy [Letter to the editor]. *Science, 282*, 1047.
Comings, D.E., Muhleman, D., Gade, R., et al. (1997). Cannabinoid receptor gene *(CNR1)*: Association with IV drug use. *Molecular Psychiatry, 2*, 161-168.
Costa, P.T., & McCrae, R.R. (1988). From catalog to classification: Murray's needs and the five-factor model. *Journal of Personality and Social Psychology, 55*, 258-265.
Dawson, E. (1995). Identification of a polymorphic triplet repeat marker for the brain cannabinoid receptor gene: Use in linkage and association studies. *Psychiatric Genetics, 5*, 550.
Devane, W.A., Hanus, L., Breuer, A., et al. (1992). Isolation and structure of a brain constituent that binds to the cannabinoid receptor. *Science, 258*, 1946-1949.
DeVry, J.M., Schreiber, R., Glaser, T., & Traber, J. (1992). Behavioral pharmacology of 5-HT1A agonists: Animal models of anxiety and depression. In S.M. Stahl, M. Gastpar, J.M. Keppel Hesselink, & J. Traber (Eds), *Serotonin IA Receptors in Depression and Anxiety* (pp. 55-81). New York: Raven Press.
Duffy, J.D. (1997). The neural substrates of emotion. *Psychiatric Annals, 27*, 24-29.
Gazzaniga, M. (1985). *The Social Brain*. New York: Basic Books.
Heils, A., Teufel, A., Petri, S., Stöber, G., Riederer, P., Bengel, D., & Lesch, K.P. (1996). Allelic variation of human serotonin transporter gene expression. *Journal of Neurochemistry, 66*, 2621-2624.
Insel, T.R. (1997). A neurobiological basis of social attachment. *American Journal of Psychiatry, 154*, 726-735.
Jackson, D.N. (1984). *Personality Research Form Manual*. Port Huron, MI: Research Psychologist Press.
Jackson, D.N., Paunonen, S.V., Fraboni, M., & Goffin, R.D. (1996). A five-factor versus six-factor model of personality structure. *Personality and Individual Differences, 20*, 33-45.
Jöreskog, K.G., & Sörbom, D. (1996). *LISREL 8: User's Reference Guide*. Chicago, IL: Scientific Software International.
Kalivas, P.W., & Barnes, C.D. (1993). *Limbic Motor Circuits and Neuropsychiatry*. Boca Raton, FL: CRC Press.
LeDoux, J. (1996). *The Emotional Brain*. New York: Simon & Schuster.

Lesch, K.P., Balling, U., Gross, J., Strauss, K., Wolozin, B.L., Murphy, D.L., & Riederer, P. (1994). Organization of the human serotonin transporter gene. *Journal of Neural Transmission: Genetics Section, 95*, 157-162.

Lesch, K., Bengel, D., Heils, A., Sabol, S.Z., et al. (1996). Association of anxiety-related traits with a polymorphism in the serotonin transporter gene regulatory region. *Science, 274*, 1527-1531.

MacDonald, K. (1991). A perspective on Darwinian psychology: The importance of domain-general mechanisms, plasticity, and individual differences. *Ethology and Sociobiology, 63*, 753-773.

MacLean, P. (1990). *The Triune Brain in Evolution.* New York: Plenum Press.

Matsuda, L.A., Lolait, S.J., Brownstein, M.J., Young, A.C., & Bonner, T.I. (1990). Structure of a cannabinoid receptor and functional expression of the cloned cDNA. *Nature, 346*, 561-564.

McClelland, D.C., Koestner, R., & Weinberger, J. (1989). How do self-attributed and implicit motives differ? *Psychological Review, 96*, 690-702.

McCrae, R.R., & Costa, P.T., Jr. (1997). Personality trait structure as a human universal. *American Psychologist, 52*, 509-516.

Michelini, S., Urbanek, M., Dean, M., & Goldman, D. (1995). Polymorphism and genetic mapping of the human oxytocin receptor gene on chromosome 3. *American Journal of Medical Genetics (Neuropsychiatric Genetics), 60*, 183-187.

Miller, W.B. (1981). *The Psychology of Reproduction.* Springfield, VA: National Technical Information Service.

Miller, W.B. (1986). Proception: An important fertility behavior. *Demography, 23*, 579-594.

Miller, W.B. (1992). Personality traits and developmental experiences as antecedents of childbearing motivation. *Demography, 29*, 265-285.

Miller, W.B. (1994). Childbearing motivations, desires, and intentions: A theoretical framework. *Genetic, Social, and General Psychology Monographs, 120*, 223-258.

Miller, W.B. (1995). Childbearing motivation and its measurement. *Journal of Biosocial Science, 27*, 473-487.

Miller, W.B., & Pasta, D.J. (1993). Motivational and nonmotivational determinants of child-number desires. *Population and Environment: A Journal of Interdisciplinary Studies, 15*, 113-138.

Miller, W.B., & Pasta, D.J. (1994). The psychology of child timing: A measurement instrument and a model. *Journal of Applied Social Psychology, 24*, 218-250.

Miller, W.B., & Pasta, D.J. (1995a). Behavioral intentions: Which ones predict fertility behavior in married couples? *Journal of Applied Social Psychology, 25*, 530-555.

Miller, W.B., & Pasta, D.J. (1995b). How does childbearing affect fertility motivations and desires? *Social Biology, 42*, 185-198.

Miller, W.B., & Pasta, D.J. (2000). Early family environment, reproductive strategy, and contraceptive behavior: Testing a genetic hypothesis. In J.L. Rodgers, D.C. Rowe, & W.B. Miller (Eds.), *Genetic Influences on Human Fertility and Sexuality.* Norwell, MA: Kluwer.

Miller, W.B., Pasta, D.J., MacMurray, J., Chiu, C., Wu, S., & Comings, D.E. (1999). Genetic influences on childbearing motivation: A theoretical framework and some empirical evidence. In L.J. Severy & W.B. Miller (Eds.), *Advances in Population: Psychosocial Perspectives*, Vol. 3 (pp. 53-102). London: Jessica Kingsley.

Miller, W.B., Turchi, B.A., & Pasta, D.J. (2000). An integrative model of the microeconomic, social-normative, and psychological determinants of fertility demand. Submitted.

Navarro, M., Fernández-Ruiz, J.J., de Miguel, R., et al. (1993). An acute dose of delta-9-tetrahydrocannabinol affects behavioral and neurochemical indices of mesolimbic dopaminergic activity. *Behavioral Brain Research, 57*, 37-46.

Panksepp, J. (1998). *Affective Neuroscience: The Foundations of Human and Animal Emotions.* New York: Oxford University Press.

Pryce, C.R. (1992). A comparative systems model of the regulation of maternal motivation in mammals. *Animal Behavior, 43*, 417-441.

Rothbart, M.K., Derryberry, D., & Posner, M.I. (1994). A psychobiological approach to the development of temperament. In J.E. Bates & T .D. Wachs (Eds.), *Temperament: Individual Differences at the Interface of Biology and Behavior.* Washington, D.C.: American Psychological Association.

Scott, J.P. (1958*). Animal Behavior.* Chicago: University of Chicago Press.

Simmons, C.F., Jr., Clancy, T.E., Quan, R., & Knoll, J.H.M. (1995). The oxytocin receptor gene (OXTR) localizes to human chromosome 3p25 by fluorescence in situ hybridization and PCR analysis of somatic cell hybrids. *Genomics, 26*, 623-625.

Tinbergen, N. (1951*). The Study of Instinct.* New York: Oxford University Press.

Tooby, J. & Cosmides, L. (1992). The psychological foundations of culture. In J.H. Barkow, L. Cosmides, & J. Tooby (Eds.), *The Adapted Mind* (pp. 19-136). New York: Oxford University Press.

Udry, J.R., Morris, N.M., & Kovenock, J. (1995). Androgen effects on women's gendered behavior. *Journal of Biosocial Science, 27*, 359-368.

Wilson, E.O. (1998). *Consilience: The Unity of Knowledge.* New York: Alfred A. Knopf.

Winter, D.G., John, O.P., Stewart, A.J., Klohnen, E.C., & Duncan, L.E. (1998). Traits and motives: Toward an integration of two traditions in personality research. *Psychological Review, 105*, 230-250.

Chapter 3

GENETIC INFLUENCES ON FERTILITY BEHAVIOR: FINDINGS FROM A DANISH TWIN STUDY, 1910-1923

Hans-Peter Kohler and Kaare Christensen

Key words: fertility, heritability, twins, dominance

Abstract: Whereas most research focuses on how variations in socioeconomic conditions contribute to the diversity in fertility behavior within a population, this paper analyzes the fertility of Danish twins in order to find possible genetic influences on an individual's fertility decisions. The analyses reveal that male monozygotic twins have a significantly higher correlation of completed fertility than their dizygotic counterparts, which suggests a (dominant) genetic influence on male fertility behavior. For females, an important gene-environment interaction emerges: the difference in the correlation of completed fertility between mono- and dizygotic twins changes over time. Only for later cohorts is there evidence for a greater similarity of completed fertility among female monozygotic twins as compared to dizygotic twins. This means that the genetic influence on the fertility of females increases over the sample period.

1.   INTRODUCTION

Although cohort fertility has been declining in many developed countries, fertility patterns have not been converging. In the United States, for instance, the completed fertility of whites declined from 3.08 for the cohort born 1930-35 to 1.93 for the cohort born 1945-50 (-37%).[1] At the same time, fertility behavior became more diverse and the coefficient of variation of women's completed fertility increased from 0.65 to 0.71 (+10%) for these cohorts. Whereas most research focuses on how variations in socioeconomic conditions contribute to this diversity in fertility behavior (Becker, 1981; Coleman, 1996), this paper analyzes the fertility of Danish twins in order to find possible genetic influences on an individual's fertility decisions.

The analyses reveal that male monozygotic twins have a significantly higher correlation of completed fertility than their dizygotic counterparts. For females, an important gene-environment interaction emerges: the difference in the correlation of completed fertility between mono- and dizygotic twins changes over time. Only for later cohorts is there evidence for a greater similarity of completed fertility among female monozygotic twins as compared to dizygotic twins.

2.   THE DANISH TWIN DATA

The sample consists of 674 same-sex twin pairs from the Danish Twin Registry who were born between 1900 and 1923 and who participated in the LSADT surveys in 1995-97 (See McGue & Christensen, 1997 for a description of the survey and the data). The available fertility measures in the data include the variables measuring whether an individual had "at least one child" and the "total number of biological children" (completed fertility). In addition we know whether a twin was ever married, which is an important proximate determinant of fertility. Summary statistics for these variables are reported in Table 1 along with a more detailed description of the composition of the twin sample.

In order to compare the fertility of the twins in the sample to the Danish population, Figure 1 displays the cohort fertility rate for the Danish population, the Danish period total fertility rate lagged 30 years, and a smoothed estimator for the cohort fertility of the twins. As can be seen in Figure 1, the twins belong to the first cohorts after the demographic transition in Denmark, with a substantially reduced fertility close to replacement level. The decline in cohort fertility experienced by the cohorts born at the end of the 19th century was achieved mainly through the introduction of stopping behavior (i.e., conscious fertility control after the

Table 1. Summary statistics for twin sample

Birth year	1900-09		1910-19		1920-23	
	Males	Females	Males	Females	Males	Females
Number of twin pairs	14	47	119	275	84	135
Percentage of monozygotic pairs	0.36	0.38	0.41	0.40	0.41	0.36
Proportion of twins who have ever married	0.96	0.91	0.90	0.89	0.91	0.89
Have at least one child	0.89	0.80	0.85	0.81	0.83	0.81
Mean (SD) of number of biological children[a]	3.00 (2.21)	2.34 (1.98)	2.38 (1.61)	2.27 (1.83)	2.56 (1.84)	2.23 (1.76)
Years of elementary Education	6.63 (1.69)	7.05 (1.80)	7.14 (1.66)	6.92 (1.46)	7.10 (1.50)	7.10 (1.42)

In order to compare the fertility of the twins in the sample to the Danish population, Figure 1 displays the cohort fertility rate for the Danish population, the Danish period total fertility rate lagged 30 years.

desired number of children has been achieved) within marriage. The reasons for this decline in fertility are often attributed to socioeconomic changes occurring in the late 19th century, and also to processes of ideational diffusion and cultural evolution (Coale & Watkins, 1986; Lesthaeghe, 1983; Turke, 1989). Although the twin fertility slightly exceeds the fertility of the general population in Figure 1, the twins in the sample share the important characteristic of consciously controlled fertility levels, which is typical for the birth cohorts born after the beginning of this century.

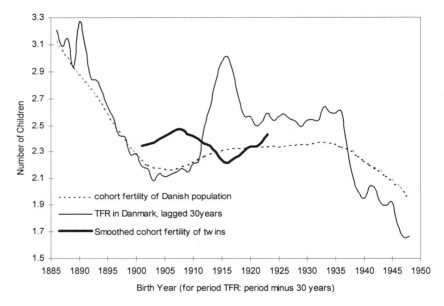

Figure 1: Cohort and period fertility trends in the Danish populations
1885-1950 compared to the fertility of the twin population.

A second aspect of fertility, which is not reflected in Figure 1, is the variation in fertility within cohorts. For the United States, historical census data allow us to analyze the parity distribution of white birth cohorts since the 1870s.[2] Although the variance in completed fertility has been declining for all cohorts since 1870, the coefficient of variation, which accounts for changing fertility levels in measuring the dispersion of completed fertility, behaves quite differently. The coefficient of variation reached a maximum for cohorts born at the beginning of this century, who had very low fertility, and it declined again during the increase in fertility that led to the baby boom. Cohorts born between 1940 and 1950 exhibited once again an increasing amount of diversity (as measured by the coefficient of variation) in their completed fertility. Hence, declining fertility during this century has not necessarily implied a convergence of fertility behaviors; low fertility may even imply a greater degree of heterogeneity in fertility choices within a population.

In this paper we investigate whether potential genetic differences in fertility behavior contribute to the diversity in observed fertility levels within cohorts, in addition to the socioeconomic and cultural differences within a population that may lead to differential fertility. The existence of such effects is not a priori self-evident. In particular, Fisher's Fundamental Theorem of Natural Selection (Fisher, 1930) suggests that in a population subject to Malthusian constraints there is relatively little genetic variation in a fitness trait such as fertility behavior.[3] However, the demographic transition dramatically altered the context of fertility behavior compared to the historical situation. Once conscious fertility control became commonplace, previously latent genetic influences on fertility behavior could potentially reveal themselves in differential fertility outcomes.

3. FERTILITY BEHAVIOR AND GENETIC INFLUENCES

3.1 Standard Biometrical Results

Table 2 gives the estimated influences of additive genetic, dominant genetic, shared environment and non-shared environment effects for the variables "ever married", "at least one child" and "number of biological children" on the basis of standard biometric models for twins (Neale & Cardon, 1992). Models with either an additive genetic effect (ACE) or a dominance genetic effect (DCE) were estimated. The results indicate in both specifications a slightly higher heritability for males than for females in both marriage and fertility behavior. However, the results are not statistically significant except for the male DCE model with a dominance genetic effect. In addition, the models in Table 2 suggest that shared environmental influences may be more important for females than for males.

Whereas these standard biometrical models yield a pattern of inheritance similar to that inferred from the econometric models below, these biometric methods are not completely satisfactory for the analysis of fertility with the available Danish twin data. First, biometrical analyses result in

Table 2. Biometrical estimation of degree of heritability (h^2), dominance effect (d^2), shared environment (c^2), non-shared environment (e^2).

Dependent Variable	Model	Males				Females			
		h^2	d^2	c^2	e^2	h^2	d^2	c^2	e^2
Ever married	ACE	75 (0-95)	-	0 (0-53)	24 (5-64)	64 (6-93)	-	17 (0-64)	18 (6-38)
	DCE	-	79 (8-96)	0 (0-53)	21 (4-59)	-	43 (4-83)	39 (2-68)	18 (7-38)
At least one child[a]	ACE	52 (0-71)	-	0 (0-43)	48 (29-73)	41 (0-72)	-	4 (0-53)	56 (28-90)
	DCE	-	46 (26-73)	12 (0-43)	42 (0-74)	-	27 (0-53)	17 (26-90)	56 (0-73)
Number of Children[a,b]	ACE	39 (3-58)	-	0 (0-23)	62 (42-85)	11 (0-49)	-	24 (0-42)	64 (50-79)
	DCE	-	45 (12-64)	0 (0-20)	55 (36-79)	-	7 (0-35)	28 (8-42)	64 (50-79)

Notes: 95% confidence intervals are in parentheses. (a) Only biological children. (b) The number of children was categorized into three groups: 0 children, 1-2 children, 3 or more children for the analysis.

substantial standard errors due to the relatively small sample size. Second, these analyses do not reveal whether similarity in fertility is partially mediated through similarity in other socioeconomic characteristics. Most importantly, according to standard models of fertility behavior, similarity in fertility patterns could be partly explained by the similar education among twins (see e.g. Behrman & Taubman, 1989). In order to learn about genetic influences on fertility, therefore, we are particularly interested in the influences net of genetic influences on educational attainment, which is not reflected in Table 2. Third, biometrical analyses do not allow us to detect changes in the level of cohort fertility and in the extent of genetic influences over time, unless the statistical power of the analysis is further reduced by grouping the sample into smaller categories.

3.2 An Econometric Approach

In order to overcome these limitations, we estimate a bivariate probit model for the variables "ever married" or "at least one child", and a bivariate ordered probit model for the variable "number of (biological) children" (see e.g. Lesaffre & Molenberghs, 1991, or Kohler & Rodgers, 1999 for a discussion of these models). Before presenting the results of this analysis, we motivate this approach from a microeconomic perspective in the next paragraphs. For this purpose, a simple economic model of fertility decisions is devised below.[4]

Assume that an individual derives utility from consuming the good c and from the number of children n, where for the moment we ignore the fact that the number of children can only be chosen as an integer number. In order to incorporate the theme of this paper, namely the existence of genetic influences on fertility preferences, in this model we allow for differences in individual's preferences for children. That is, we allow some individuals to have stronger preferences for children, whereas others have stronger preferences for consumption. We assume that this heterogeneity in preferences is reflected in a parameter θ of the utility function.

In order to make optimal fertility decisions individuals therefore maximize a utility function

$$U(c, n; \theta)$$

subject to a budget constraint $I = p_c c + p_n n$, where p_c denotes the costs of the consumption good c and p_n denotes the costs of children n. When $\partial^2 U / \partial n \partial \theta > 0$, then increases in the level θ are associated with a higher valuation of children in the utility function. We also assume that the income I of an individual depends partly on individual characteristics X, and additionally on random effects v reflecting labor market conditions, etc. We assume that v is known to the individual when the fertility decision is made. The optimal number of children n^* in this example is therefore a function of three factors: (a) the cost of children p_n and of the consumption good p_c, (b) the individual characteristics X and the random shock v which determine income, and (c) the parameter θ, which reflects preferences for children. Formally, this demand for children is written as

$$n^* = d_n(\ p_c\ ,\ p_n\ ,\ X,\ v,\ \theta\ ) \qquad\qquad (1)$$

where $d_n(.)$ is the demand function for children. For the econometric estimation we specify this demand function as a linear function of the observable characteristics X as

$$n^* = X\ \beta + \varepsilon. \qquad\qquad (2)$$

In twin studies we utilize the similarity of the outcome variable, which is the number of children in the above example, in order to infer possible genetic influences on behavior. When n_1^* and n_2^* denote the fertility chosen by the first and second twin, then a correlation between n_1^* and n_2^* can arise due to two different effects. The fertility among twins could be similar due to a similarity in individual characteristics X, which affect the budget constraint, or more generally, the costs of children to an individual. Alternatively, the twins can exhibit similar fertility due to a correlation in the random term ε, which comprises the correlation of the preference parameter θ between twins, as well as the correlation of other unobserved characteristics. The exact econometric identification of a correlation in the preference parameter θ is only possible under very strong assumptions and is not pursued here. Nevertheless, it is important to note that genetic influences on fertility preferences reveal themselves in a persisting correlation in the fertility of twins, even after the characteristics X are controlled for.

3.3   Estimation

In order to infer genetic influences on fertility behavior we utilize the fertility related variables "ever married", "at least one child", and "number of biological children", which are observed for each twin in the dataset. Since the former two variables are dichotomous and the number of children is an ordered categorical variable, we estimate a bivariate (ordered) probit model to infer the correlation between twins in the random term ε in equation 2.

In these bivariate (ordered) probit models each twin pair constitutes one pair of observations. Similar to equation 2, the dependent variables are modelled as a function of observable characteristics X, which include birth year, birth order, sex, and in some models the years of primary education.

Formally, the bivariate (ordered) probit model is specified as follows: n_{ij}^* is a latent scalar measuring the propensity to have children (or the propensity to marry) with $n_{ij}^* = x_{ij}\beta + \varepsilon_{ij}$, where $j = 1, \ldots, N$ denotes twin pairs, $i = 1, 2$ denotes the particular twin within twin pairs, x_{ij} is a $1 \times K_1$ vector of covariates, and ε_{ij} is a normally distributed random term which is independent across twin pairs but correlated within pairs. The observed variable n_{ij} equals $0, 1, \ldots, \omega$ if the latent variable n_{ij}^* falls into the intervals $(-\infty, c_1], \ldots, (c_w, \infty)$, where $c_1, \ldots, c_\omega$ are estimated cut points and $\omega + 1$ is the number of categories. The bivariate probit model is a special case with only two categories and $\omega = 1$. The correlation of the random term across twins is modeled as $\rho_j = z_j \delta$, where z_j is a $1 \times K_2$ vector of covariates describing characteristics of the twin pair. The coefficients in β, and therefore the influence of the characteristics x_{ij} on the dependent variable n_{ij}, is assumed to be independent of the zygosity of twins. Moreover, the coefficients of the explanatory variables are constrained to be equal across twins. The coefficients β, ρ are estimated via a maximum likelihood estimation (see e.g. Kohler & Rodgers, 1999; Lesaffre & Molenberghs, 1991).

The correlation coefficient ρ of the random term reflects the correlation in the latent propensity to have children (propensity to marry) after controlling for the characteristics x_{ij}. Differences in this correlation between mono- and dizygotic twins allow us to infer that there are genetic influences on the propensity to marry or have children. In order to obtain estimates of this correlation for all sex and zygosity combinations, the correlation is modeled as $\rho = \delta_1 d_M + \delta_2 d_{MM} + \delta_3 d_F + \delta_4 d_{FM}$, where d_M, d_{MM}, d_F, d_{FM} are dummy variables for male, male-monozygotic, female, female-monozygotic twin pairs respectively. The coefficient δ_1 (δ_3) gives the correlation of the random term for male (female) dizygotic twins, whereas the coefficient δ_2 (δ_4) measures the difference in the correlation between male (female) monozygotic twins and their dizygotic counterparts. In some models an additional interaction term $\delta_5 * sex * birth year * monozygotic$ is included in the equation for ρ in order to detect changes over time.

Table 3 shows the results of the bivariate probit estimations for the variables 'ever married' and 'at least one child'. In Models 1 and 3, only birth year and sex are used as explanatory variables, whereas Models 2 and 4 also include the birth order and years of primary education. The effect of education on the dependent variable is allowed to differ for males and females. Birth order is included in order to control for differential treatment of the twins according to the birth order. Years of primary education controls for the more similar educational achievement of monozygotic twins. Specifically, the correlation in the years of primary education is 0.67 and 0.10 for male mono- and male dizygotic twins, or 0.41 and 0.19 for female mono- and female dizygotic twins respectively. The years of primary education are included among the explanatory variables in order to control for the extent to which the similarity in fertility behavior between twins is mediated through a similarity in educational achievement, especially for monozygotic twins. Years of secondary education were not included among the explanatory variables since it is likely to be correlated with unobserved characteristics related to fertility behavior/preferences, and the direction of causality between fertility and education is substantially less clear when higher education is considered.

Model 1 reveals that the propensity to have ever married does not vary significantly with sex or birth year. The random term ε, which reflects the propensity to marry net of the influence of the explanatory variables, is not significantly correlated for male DZ twins. Male MZ twins, on the other hand, differ significantly from their DZ counterparts ($\delta_2 = 0.61$) and exhibit a correlation of $\delta_1 + \delta_2 = 0.80$ in their latent propensity to have ever married. The results for females differ from those for males in that female DZ twins exhibit a significant correlation in the random term ε ($\delta_3 = 0.5$), which suggests a common environment effect for females. MZ female twins are, like their male counterparts, more similar in their marriage behavior than their DZ counterparts ($\delta_4 = 0.32$), although the MZ–DZ difference is smaller for females than for males.

Interestingly, in Model 2, which controls for additional explanatory variables, marriage behavior is associated with a birth order effect: the first born twin within a twin pair is more likely to have ever married than the second born twin. This finding potentially points to some differential parental treatment of the twins according to birth order. With respect to the correlation pattern in the random term ε, Model 2 yields findings similar to the previous estimation. Nevertheless, controlling for the additional explanatory variables increases the correlation in the random term for male and female DZ twins (δ_1 and δ_3), and it decreases the difference in correlation between MZ and DZ twins (δ_2 and δ_3). This finding supports the argument that the greater similarity in marriage behavior of MZ twins as compared to DZ twins is in part mediated through a greater similarity in other characteristics, such as education for instance.

Model 3 reports the bivariate probit analysis for 'at least one child', using only birth year and sex as explanatory variables. Similar to the previous models, there is no significant difference in the probability of having at least one child due to sex or age of the twins. The correlation of the random term reveals that male DZ twins do not exhibit a significant correlation in the

Table 3. Results of bivariate probit estimation for the variables "ever married" and "at least one child"

Dependent variable Method	Ever married bivariate probit[b]		At least one child[a] bivariate probit[b]	
	Model 1	Model 2	Model 3	Model 4
birth year	-0.7E-3	0.001	0.009	0.009
	(0.010)	(0.010)	(0.008)	(0.009)
female	-0.071	-0.269	-0.121	0.626
	(0.099)	(0.462)	(0.087)	(0.412)
born first	-	0.194	-	-0.025
		(0.092)**		(0.081)
male*years of	-	-0.022	-	0.010
primary education[c]		(0.053)		(0.046)
female*years of	-	0.003	-	-0.096
primary education[c]		(0.035)		(0.032)***
correlation	0.178	0.246	-0.132	-0.083
male (δ_1)[d]	(0.237)	(0.239)	(0.217)	(0.227)
male*				
monozygotic (δ_2)[d]	0.617	0.510	0.499	0.452
	(0.273)**	(0.279)*	(0.293)*	(0.300)
female (δ_3)	0.497	0.534	0.367	0.363
	(0.126)***	(0.125)***	(0.108)***	(0.109)***
female*				
monozygotic (δ_4)	0.322	0.275	0.207	0.146
	(0.148)**	(0.151)*	(0.156)	(0.164)
# of twin pairs[e]	670	661	668	660

p-values: $*p < 0.1$; $**p < 0.05$; $***p < 0.01$.

Notes: (a) Only biological children. (b) The constant term is not reported. (c) Years of secondary education were not included among the explanatory variables since it is likely to be correlated with unobserved characteristics related to fertility behavior/preferences. (d) A Wald test rejects the null hypothesis of H_o: $\delta_1 + \delta_2 = 0$ for all models at a 94% or higher confidence level. (e) Some twin pairs were excluded from the analyses due to missing information.

propensity to have children. Their male MZ counterparts, on the other hand, have a significant correlation of about $\delta_1 + \delta_2 = 0.36$, which is significantly higher than the correlation of the male DZ twins. Interestingly, the findings for females differ quite substantially from the above results. For females the propensity for having at least one child is significantly correlated across twins, but female MZ twins do not differ significantly from DZ twin pairs. Model 4 controls additionally for the birth order and the years of primary education. The strongest influence of these covariates on the probability of having children occurs with respect to female education. Increases in female education tend to reduce the probability of having children. The pattern of correlation in the error term largely coincides with Model 3, with the relevant difference being a reduction in δ_2 and δ_4. This again indicates that part of the greater similarity among MZ twins is mediated through a more similar educational achievement.

Models 5 and 6 in Table 4 report the corresponding bivariate ordered probit model for the number of biological children. The results are similar to the findings reported above for the variable 'at least one child': Male DZ twins exhibit no significant correlation in their propensity for having children (δ_1 is not significantly different from zero), whereas male MZ twins are significantly more similar in their fertility behavior than their DZ counterparts (both δ_2 and $\delta_1 + \delta_2$ are significantly different from zero). Female twins generally exhibit a significant correlation in completed fertility (δ_3), but there is no difference between the correlation of MZ and DZ twins (δ_4 is close to zero). Controlling for education slightly reduces the observed correlations, but it does not change the overall pattern.

Figure 2 provides a visual representation of the estimated correlation pattern of Model 5 in Table 4. The numbers in each figure give the 7×7 table of the number of children by a twin pair for all sex-zygosity combinations. Superimposed on these 7×7 tables are isoquants of the bivariate normal distribution of the unobserved propensity n_{ij}^* to have children[5] for the respective sex and zygosity combination. The dotted lines represent the estimated cut points $c_1, ..., c_6$ which translate the unobserved propensity for having children n_{ij}^* into the observed number of children of each twin in the bivariate ordered probit model.

The top panel reveals again the finding that male dizygotic twins do not exhibit a correlation in their propensity for having children, whereas their male MZ counterparts do exhibit a significant correlation, which leads to a greater similarity in the fertility behavior within twin pairs. The 7×7 table with the observed fertility of twin pairs is more oriented along the diagonal for MZ twins, and similarly is the distribution of the latent propensity for

Table 4. Results of bivariate probit estimation for "number of biological children"

Dependent variable Method	Number of children[a] bivariate ordered probit[b]		
	Model 5	Model 6	Model 7
birth year	0.004	0.003	0.002
	(0.006)	(0.006)	(0.006)
female	-0.161	-0.023	-0.013
	(0.060)***	(0.273)	(0.273)
born first	-	0.045	0.051
		(0.056)	(0.056)
male*years of	-	-0.0407	-0.041
primary education[c]		(0.030)	(0.030)
female*years of	-	-0.061	-0.062
primary education[c]		(0.023)***	(0.023)***
correlation	-0.039	0.038	0.040
male (δ_1)[d]	(0.098)	(0.102)	(0.102)
male*monozygotic (δ_2)[d]	0.511	0.439	0.436
	(0.133)***	(0.136)***	(0.136)***
female (δ_3)	0.271	0.257	0.256
	(0.059)***	(0.059)***	(0.059)***
female*			
monozygotic (δ_4)	0.025	0.005	-0.685
	(0.092)	(0.094)	(0.284)***
female*birthyear*			0.042
monozygotic (δ_5)[e]			(0.015)***
number of twin pairs[e]	668	660	660

p-values: *p < 0.1; **p < 0.05; ***p < 0.01.

Notes: (a) Only biological children. (b) The categories are zero children, one child, . . ., five children, six or more children; the Cutpoints for the categories are not reported. (c) Years of secondary education were not Included among the explanatory variables since it is likely to be correlated with   unobserved characteristics related to fertility behavior/preferences. (d) A Wald test rejects the null Hypothesis of H_o: $b_1 + b_2 = 0$ for all models at a 94% or higher confidence level. (e) The estimated coefficient b_5 and the linear interaction with birth year do not lead to invalid correlations with $|\rho| > 1$ for the birth years 1900-23, which are represented in the data. A wald test of the null-hypothesis H_o: b_4 + female*birth year*$b_5 = 0$ in model 5 is rejected at the 94% or higher confidence level for cohorts born after 1920. (f) Some twin pairs were excluded from the analyses due to missing information.

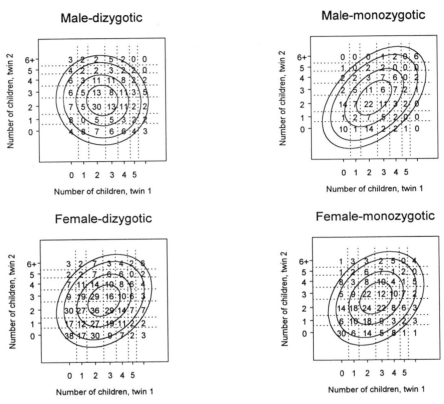

Figure 2: Visual representation of bivariate ordered probit model (model 5 in Table 4): The numbers in each figure give the 7×7 table of the number of children by a twin pair for all sex-zygosity combinations. Superimposed on this 7×7 table are the isoquants of the bivariate normal distribution of the unobserved propensity n_{ij}^* for having children (The calculations were based on the birth year 1916, but the influence of birth year on the number of children is quite negligible in the estimated results). The dotted lines represent the estimated cut points $c_1, \ldots, c_6$ which translate the unobserved propensity for having children n_{ij}^* into the observed number of children of each twin in the bivariate ordered probit model. The isoquants are chosen to delimit the 50%, 75%, 90% and 95% intervals of the marginal distributions of n_{ij}^*, measured at the center of the distribution.

having children. For the female fertility, which is drawn in the bottom panel of Figure 2, DZ twins exhibit a correlation in their fertility behavior which was absent in the male twin pairs. However, the significant difference in correlation between male MZ and DZ twins is not found for females: female MZ and DZ twins exhibit an almost identical correlation in their propensity for having children, and the distribution of n_{ij}^* is almost identical in the left and right figure in the bottom panel of Figure 2.

In order to investigate the absence of a difference in correlation between female MZ and DZ twins further, model 6 in Table 4 is augmented with an interaction term $\delta_5 * sex * monozygotic * birth year$. This interaction term measures whether the difference in correlation between MZ and DZ twins changes over time. For males, no significant change in the correlation over time is found and the results are therefore not reported in Table 4. For female twins, however, an interesting time pattern emerges, which is reported in right-most column of Table 4 (model 7): the positive coefficient δ_5 implies that monozygotic twins become more similar over time. In particular, in later cohorts, female MZ twins exhibit a significantly higher correlation than DZ twins. For later cohorts, the shape of the distribution of n_{ij}^* in the bottom right graph of Figure 2 becomes more and more oriented along the diagonal; for cohorts born after 1920 it approaches the shape of the distribution for male MZ twins, which is depicted in the top right graph of Figure 2.

The time dependence of the correlation in the propensity for having children suggests a gene-environment interaction: the genetic influences on fertility depend on environmental conditions, and changes in the socioeconomic conditions may determine whether the genetic influences result in observable correlations in behavior among twins.

4. DISCUSSION AND CONCLUSION

Interpreting the findings in Tables 3 and 4 in terms of genetic influences on marriage and fertility behavior, the results indicate a substantial genetic influence for males. The results reveal a modest correlation in the propensity for marrying among male DZ twins, and a significant difference in this propensity between male MZ and DZ twins. In terms of fertility, neither the variable `at least one child' nor the variable `number of biological children' reveals a correlation in the propensity for having children for male DZ twins. At the same time, male MZ twins exhibit a significant correlation in this propensity, and they differ significantly from their DZ counterparts. This difference in correlation between male MZ and DZ twins is quite substantial (δ_2 ranges from 0.43 to 0.51) and this suggests relevant genetic influences on the fertility of males. In particular, the absence of correlation among DZ twins suggests that this influence is primarily due to a dominance effect, and this interpretation is consistent with the results of the biometrical estimates in Table 2. The similarity of fertility behavior within male MZ twin pairs is partly mediated through a similarity of schooling attainment. A quite substantial part of this correlation in fertility of MZ twins remains however even after one controls for years of education. Correlated educational attainment, therefore, does not seem to be the primary reason for the similarity in the fertility behavior of male MZ twins.

The analysis of the marriage behavior of females yields a relevant shared environment effect (reflected in the significant correlation between DZ twins, exceeding half of the MZ correlation) and a genetic influence that reveals itself in a significant difference in the correlation of female MZ and DZ twins. The parameter estimates suggest that this genetic influence for females is somewhat weaker than for males. With respect to fertility, the above analyses find a significant correlation in the propensity for having children between female DZ twins, but the correlations of female MZ and DZ twins do not differ significantly in Models 3 through 6. This finding suggests a greater relevance of shared and non-shared environment effects for female fertility, which is also indicated by the biometrical analyses in Table 2.

Further investigation of the female pattern reveals the presence of a gene-environment interaction. In Model 7 (Table 4) the correlation in the propensity for having children is allowed to vary over time. Whereas for males no effect is found, the significant coefficient δ_5 for the interaction term *female * monozygotic * birth year* reveals that the difference in correlation between female MZ and DZ twins does in fact vary over time. Hence the genetic influence on fertility behavior changes with birth cohorts. Only for latter cohorts the fertility behavior of monozygotic twins is more similar than that of dizygotic female twins, and hence the genetic influence on fertility behavior is stronger for female twins who are born late in the period 1900-23. This time-dependence of heritability patterns is also found in Kohler, Rodgers, and Christensen (1999), who have documented substantial changes in genetic and shared environmental influences on fertility for cohorts born during 1870–1910 and 1953–64.

We propose that the above findings are due to genetic influences on fertility preferences that partly explain the diversity in fertility behavior within a population. The investigated cohorts clearly controlled fertility, as the average number of children is 2.26 for males and 2.49 for females in the sample. In addition, the results for 'at least one child' and 'number of biological children' correspond closely, and the results are robust even after education is controlled. This suggests that conscious decisions about fertility levels, rather than effects operating through biological fecundity, underlie the findings.[6]

This interpretation is also consistent with a potential explanation for the peculiar female pattern. Since fertility is generally a household decision reflecting the preferences of husband and wife, correlated fertility preferences among twins reveal themselves in the number of children a couple has, provided that the household decision includes the preferences of the twin. Whereas the females in early cohorts experienced most of their reproductive years before WWII, the reproductive years of the later cohorts extended into the 1960s. An important development during and after the war is its influence towards an increased household bargaining power for females. Correlated fertility preferences between female twins may not have revealed themselves in an observed similarity of fertility when female influence on household decisions was low, but they did as the female influence on household decisions increased after WWII.

ACKNOWLEDGMENTS

We are grateful to James W. Vaupel for suggesting this collaborative project and for helping provide support for it. We are also grateful for suggestions and comments by Joseph L. Rodgers, Mike Murphy, Kenneth Wachter and the participants of the NIH conference, Genetic Influences on Fertility Related Processes. We gratefully acknowledge the support we have received from the Max Planck Institute and the National Institute on Aging (research grant NIA-PO1-AG08761) and from the Danish Interdisciplinary Research Council. The activities of the Danish Center for Demographic Research are funded by a grant from the Danish National Research Foundation.

REFERENCES

Becker, G. S. (1981). *A Treatise on the Family*. Cambridge: Harvard University Press.

Behrman, J. R. & Taubman, P. (1989). Is schooling "mostly in the genes"? Nature-nurture ecomposition using data on relatives. *Journal of Political Economy, 97*, 1425–1446.

Christensen, K., Basso, O., Kyvik, K. O., Juul, S., Boldsen, J., Vaupel, J. W., & Olsen J. (1998). Fecundability of female twins. *Epidemiology, 9*, 189–192.

Cigno, A. (1991). Economics of the Family. Oxford: Clarendon Press.

Coale, A. J. & Watkins, S. C. (1986). *The Decline of Fertility in Europe*. Princeton: Princeton University Press.

Coleman, D. (1996). *Europe's Population in the 1990s*. Oxford: Oxford University Press.

Fisher, R. A. (1930). *The Genetical Theory of Natural Selection*. Oxford: Clarendon Press.

Kohler, H.-P. & Rodgers, J. L. (1999). Df-like analyses of binary, ordered and censored variables using Probit and Tobit approaches. *Behavior Genetics, 29*, 221-232..

Kohler, H.-P., Rodgers, J. L., & Christensen, K. (1999). Is fertility behavior in our genes: Findings from a Danish twin study. *Population and Development Review, 25*, 253–288.

Lesaffre, E. & Molenberghs, G. (1991). Multivariate probit analysis - a neglected procedure in medical statistics. *Statistics in Medicine, 10*, 1391–1403.

Lesthaeghe, R. (1983). A century of demographic and cultural change in Western Europe: An exploration of underlying dimensions. *Population and Development Review, 9*, 411–435.

McGue, M. & Christensen, K. (1997). Genetic and environmental contributions to depression symptomatology: Evidence from Danish twins 75 years of age and older. *Journal of Abnormal Psychology, 106*, 439–448.

Murphy, M. (1998). Is the relationship between fertility of parents and children really weak? An analytic review. London: London School of Economics, Mimeo.

Neale, M. C. & Cardon, L. R. (1992). *Methodology for Genetic Studies of Twins and Families*. London: Kluwer Academic Publishers.

Ruggles, S. and M. Sobek (1997). Integrated Public Use Microdata Series: Version 2.0. Minneapolis: Historical Census Project, University of Minnesota.

Turke, P. W. (1989). Evolution and the demand for children. *Population and Development Review, 15*, 61–90.

Notes

1 The cohort fertility and standard deviation were estimated using the IPUMS census data from 1970-1990 (Ruggles and Sobek 1997).

2 Historical US census information from 1900-90 was used to calculate the parity distribution of white birth cohorts. See Ruggles and Sobek (1997) for a description of the data.

3 See e.g. Murphy (1998) for a review of the literature on parent-children correlations in fertility. This literature frequently argues that these correlations have been weak in most pre-transitional populations.

4 Since this model is mainly for illustration, a very simple formulation is chosen. For a more detailed discussion of economic fertility models and newer theoretical developments see Becker (1981) or Cigno (1991).

5 The calculations were based on a birth year of 1916, but the influence of birth year on the number of children is quite negligible in the estimated results.

6 A comprehensive analyses of fecundability among twins and singletons concluded that the fecundability of twins is comparable to that of singletons. (Christensen et al. 1998)

Chapter 4

GENETIC AND ENVIRONMENTAL INFLUENCES ON FERTILITY EXPECTATIONS AND OUTCOMES USING NLSY KINSHIP DATA

Joseph Lee Rodgers and Debby Doughty

Key words: Fisher's theorem, fertility, fertility expectations, NLSY, heritability, shared environmental influences, DF analysis

Abstract: There has been recent interest in the research literature concerning the potential for genetic influences on fertility-related behaviors. Fisher's (1930) well-known theorem suggesting that the heritability of fertility-linked behaviors must eventually disappear (e.g., Plomin, DeFries, & McClearn, 1990) runs counter to a number of empirical findings concerning sexuality and fertility behaviors. Miller has recently developed a framework (Miller et al, 1999b) that casts fertility outcomes into the bigger context of fertility desires and expectations. We draw on this framework to investigate the role of broad genetic and environmental influences on a number of fertility attitudes, and link those to fertility outcomes. Our data come from recently defined kinship structure from the National Longitudinal Survey of Youth, and thus provide a large national sample in which to investigate these issues. Our findings suggest that both fertility expectations and desires have a heritable component, and virtually no shared environmental component. However, expectations have a systematically higher level of genetic influence than outcomes. These findings are both readily interpretable within previous frameworks, and also can be used to general future research agendas.

1. INTRODUCTION

Fisher's (1930) theorem suggested that behaviors related to fertility cannot, in the long run, show a heritable component. But there is substantial empirical evidence that many such behaviors are, indeed, influenced genetically. In fact, Fisher himself, in the same source in which he published his theorem (Fisher, 1930), couldn't resist doing an empirical analysis to test for genetic influences on fertility outcomes in a three-generation British dataset. Interestingly, he found a moderate heritability in this data source, and speculated on the influence that contraception and other social factors would have on the heritability of fertility. So while Fisher's empirical analysis appeared inconsistent with the basic conclusion of his fundamental theorem, others have found support. Williams & Williams (1974) re-analyzed Fisher's original data and questioned whether there really was heritability of completed fertility in that data source. They attributed Fisher's finding of a significant heritability to his failure to account for a secular change in fertility that artifactually created the kinship correlations that led to his estimate. Imaizumi, Nei, & Furusho (1970; also see Burt, 1995) found nonsignificant heritabilities in separate analyses of father-child (h^2 =-.02) and mother-child (h^2 =.12) correlations. Also, Mealey and Segal (1993) estimated heritability in number-of-children using data from MZ and DZ twins raised apart, and found an MZ correlation of .06 (implying h^2 =.06; n=32 pairs) and a DZ correlation of r=.10 (implying h^2 =.20; n=23 pairs); neither was significant. Mealey and Seagel's sample had an average age of 38 years (sd=15) for males and 42 years (sd=11) for females; as a result their measures of number of children were somewhat censured as estimates of completed fertility. Burt (1995) presented a table showing heritability and genetic variance estimates across several other species in addition to homo sapiens. But apart from the articles above, little empirical evidence exists addressing the issue of heritability of human fertility.

On the other hand, substantial evidence supports the existence of significant genetic influences on reproductive and fertility behaviors that eventually lead to fertility outcomes. Significant heritabilities have been found on several fertility-related behavior, including sexual behavior (Dunne et al., 1997; Martin, Eaves, & Eysenck, 1977; Miller et. al., 1999a; Rodgers, Rowe, & Buster, 1999), divorce (McGue & Lykken, 1992), and parenting behaviors (Mealey & Segal, 1993; Perusse, Neale, Heath, & Eaves, 1994). Fisher's theorem applies to these measures as well as to completed fertility (at least to the extent that these fertility-related variables provide selective advantages). However, emerging theoretical work and re-interpretations of Fisher's theorem published in the biological literature have suggested reasons why these types of findings might be expected to occur (e.g., Burt,

1995; Edwards, 1994; Ewens, 1989; Houle, 1992; Lessard, 1997). In particular, environmental changes and other "perturbing forces" (see Houle, 1992) are not accounted for in Fisher's theorem.

An intriguing line of research related to this issue -- and one that directly motivates the current study -- is a series of articles by Miller and his colleagues addressing the role of genetic influences on fertility desires and expectations. Fertility researchers during the 1970's developed elaborate theoretical distinctions between the concepts of ideal, expected, desired, and intended fertility outcomes (e.g., Ryder & Westoff, 1969). While we will not attend to all the subtleties of these distinctions, we do have information available that allows some of these distinctions to be meaningful. But this is a small issue in the context of our overall treatment of the issue of whether there are genetic influences that apply to fertility planning.

In the first section of this paper, we will develop a theoretical orientation around the arguments that Miller and his colleagues have proposed to motivate the possibility of genetic influences on fertility planning and expectations. In addition, we will theoretically link these to fertility outcomes, and make some predictions about how genes might differentially affect fertility expectations and outcomes. Next, we will describe a new and unique national data source that allows investigation of the genetic and environmental influences on fertility expectations and outcomes. Next, we describe the analytic method we will use to investigate the role of genetic influences on these fertility processes. Following, we present results of an analysis of these data, and discuss the findings in the context of the theoretical framework.

2. THEORETICAL FRAMEWORK

Miller et al. (1999b) developed a broad framework that helps to motivate our work in this paper. They suggested that childbearing motivation is itself potentially under genetic control. They developed an elaborate biosocial model of this transmission process in which dopamine receptor genes play an important role (see Miller et al., 1999a, for empirical support for the role of dopamine genes in accounting for timing of first intercourse). They note that "Except for the occasional passing acknowledgment of a possible biological basis for childbearing motivation ..., very little has been said about the biological contributions to the motive forces that underlie human childbearing."

The basis of our theoretical framework is an interest in the link between the genetics of fertility expectations and fertility outcomes. Are fertility motivations, preferences, and expectations passed on genetically in a way that is strongly linked, weakly linked, or not linked at all to fertility outcomes? We examine each of these possibilities in turn. If fertility expectations are not linked at all to fertility outcomes, then by definition there should be low or no genetic variance in fertility expectations. In this case, it would appear that whatever genetic/biological components of fertility outcomes do exist have been selected through processes that are not under volitional control (at least the type of volition that would be linked to expectations and desires). This situation would be indicated by heritability of fertility expectations of around zero, with moderate heritability for fertility outcomes (the opposite of the empirical findings reviewed in the previous section) . If desires/expectations/motivations for family size are strongly related to fertility outcomes, then heritability in these fertility attitudes translate directly into heritability in fertility outcomes, and both should be moderate to high, with fertility outcomes having higher heritabilities because there are other potential sources of genetic influence on them than just fertility expectations (e.g., nonvolitional processes related to fecundity). This situation would be indicated by medium-to-large sized heritabilities for fertility expectations and fertility outcomes, with somewhat higher heritabilities for fertility outcomes (again, a result inconsistent with the empirical evidence). In this case, volitional factors would appear to explain a great deal of the variance in fertility outcomes. But neither of these extreme positions -- that there is no volitional influence, and that volition accounts for all of the variance -- seems likely, nor are they empirically supported by past work.

The middle alternative — that genetic influences on fertility expectations are imperfectly translated into fertility outcomes — seems the most plausible, for several different reasons. First, we know that small-to-moderate heritabilities for fertility-related behaviors that influence completed fertility have been found in past research. Further, low or zero heritabilities have been found for completed fertility. Thus, the combination of a number of studies suggests that the link between fertility expectations and outcomes will be weak, if it exists at all. Second, Miller et al.'s (1999b) argument gives us a reason to expect genetic influences on one important fertility attitude, childbearing motivation.

The difference between volitional influences on fertility and those that are embedded invisibly in the human's biological substrate is an extremely important distinction within our framework. Volitional influences can be reinforced and magnified by environmental influences just as pure biological influences can be. For example, a desire for a very large family translated from a mother to her daughter genetically might be realized completely independently of any social influence. But social influence from within the family can enhance the already genetically-driven desire for a large family.

By the same argument, a desire for a small family — which might interact with genetic and environmental influences on educational desires, occupational aspirations, etc. — can be further reinforced through many social mechanisms.

It appears that Fisher's theorem applies primarily to non-volitional factors that have a genetic influence on fertility. For example, if a pelvic structure emerges (for example, by mutation) that interferes with childbearing, it will quickly disappear as those women with that structure have few or no children. Or, if a genetic influence emerges that interferes with sexual functioning, selection will cause that trait to disappear in a few generations. But desire/expectation/motivation for children has a different status. As long as a strong biological tendency toward having *zero* children is avoided, this genetic goal can exist for an unlimited number of generations. The basic difference between these two situations, as it applies to Fisher's theorem, appears to be the distribution of family sizes selected for. If that distribution includes many childless individuals, the genetic variance for that trait will disappear. But suppose the distribution selected for includes no childless individuals. Under different models of selection and population size, the proportion of descendants of those with high and low family size desires may shift dramatically. But there will always be descendants of the low-family-size ancestors in the population (and thus additive genetic variance will exist on that trait, contrary to the expectations of Fisher's theorem).

This line of argument leads us to expect possible heritability for fertility desires/expectations/motivation. Further, it is easy to see how moderate to high levels of heritability for these preliminary attitudes and behaviors might be attenuated as they are passed onto fertility outcomes. This attenuation occurs because people cannot perfectly achieve their fertility goals. Random influences, partner influences, and competition from other activities of life (e.g., education and occupation) play a role as well. Thus, we propose, for fertility desires/expectations/motivations to maintain a genetic influence, they will probably require larger heritabilities than the fertility outcomes themselves.

Based on these arguments, we predict that at least some (or maybe most) of the heritable part of fertility outcomes derives from fertility desires, expectations, and motivation. Thus, we predict a moderate level of genetic influence on these fertility attitude measures, and a smaller level of genetic influence on fertility outcomes themselves. We note that this prediction is at least partially consistent with empirical findings by Mealey and Segal (1993), the only other study of which we are aware that looked at genetic variance associated with both reproductive/childbearing motivation and fertility outcomes; they found moderate heritabilities for dating (h^2 =.33 for

MZ twins), marriage (h^2 =.76 for MZ twins), and desire for children (h^2 =.33 for MZ twins), and much lower ones for number of children (h^2 =.06 for MZ twins).

The idea that fertility desires affect completed fertility also suggests a possible role for shared environmental influences as well, as well as a possible environment by gene interaction. To the extent that parents pass on to their children through socialization and parental influence desires for small, medium, or large families, these will be observed as shared environmental influences. We have little guidance from the literature to help predict whether these will exist or not.

3. METHODS

3.1 The Data

Our data come from the National Longitudinal Survey of Youth (NLSY), a survey of over 12,000 youth in 1979 who were then followed on a yearly basis until the present. These youth resided in the 3000 households that were selected in a household sample of the U.S. In each household, all youth aged 14-22 in 1979 were evaluated on a number of different dimensions. In 1979 and 1982, ideal family size was assessed. In 1979 and 1982, desired family size was also assessed. In 1979, 1982-1986, 1988, 1990, 1992, and 1994, expected family size was assessed. Further, completed family size up to that year is available for each of these time periods as well. These measures — ideal family size, desired family size, expected family size, and completed family size — form the basic dependent variables for our study.

The use of behavior genetic designs to separate genetic and environmental influences has become increasingly popular in the recent research literature. But such designs require kinship information indicating the genetic relatedness of matched kin. Unfortunately, the NLSY survey contains no explicit indicators that allow such linking. For example, the category "sibling" contains an unknown mixture of full, half, and adoptive siblings. Using a recently developed linking algorithm, however, we have used external information to infer the genetic relatedness of a large proportion of the NLSY cousins, siblings, and twins. Details of the linking algorithm are included elsewhere (e.g., Rodgers, 1996; Rodgers, Rowe, & Buster, 1999). Using this algorithm, we have classified kinship pairs into cousin pairs (with genetic coefficient R=.125), half-sibling pairs (R=.25), full sibling pairs (R=.50), same-sex twins of unknown zygocity (R=.75 on the average), and a group of sibling links in which they are clearly either half or full siblings, but for whom exact classification is impossible (for whom we define R=.375). We use these genetic indicators in the context of a behavior genetic design that predicts that more closely genetic related kin will show higher similarity on behaviors that are influenced by genes.

Our linking algorithm provided genetic indicators (R coefficients) for 3890 kinship pairs. After using the standard double entry procedure (described in the next section), we had a dataset with 7780 observations. For computing standard errors, the original sample size of 3890 pairs was used. Of course for our various analyses there were many fewer than the 7780 total observations for a given analysis because of idiosyncratic missing data patterns associated with responses to different questions and to age patterns at which certain information was obtained. Sample sizes in tables indicate the number of observations with available data for the given analysis (out of the original 7780 observation). Thus, sample sizes indicate number of unique individuals.

We use the data structure of the NLSY to define two different analyses of the NLSY youth, a cohort analysis and an age analysis. In the cohort analysis, the whole NLSY cohort is observed and studied at each of the relevant time points (1979 and 1982 for ideal family size, in nine different years for expected family size, and in the same nine different years for completed family size). A problem with this design is that siblings who differ in age in a given year have had differential opportunity to define their completed family size; the older siblings has had more years in which to have children. We account for this problem with a regression adjustment, by regressing age out of the fertility outcomes measures.

In our age analysis, we use a more direct method to handle the problem of age adjustment, and we prefer to interpret the results from this analysis because of this direct method. To age adjust we use the longitudinal structure of the data to link the kinship pairs at the same age (and different points in time, except for twins). This design has the advantage of almost exactly equating the kinship pairs on age, but the disadvantage of smaller sample sizes. We limited our analysis to settings in which at least N=1000 (500 pairs) of kinship links were available. We had that sample size for ages 20-28, resulting in nine different age analyses.

3.2   Analytic Procedures

The analytic procedure used to estimate heritability and shared environmental variance in this study was DF Analysis (DeFries and Fulker, 1985). In DF Analysis, the following regression model is fit:

$$FM1 = b0 + b1*FM2 + b2*R + b3*(FM2*R) + e, \qquad (1)$$

where FM1 is the fertility measure for the first member of the kinship pair, FM2 is the fertility measure for the second member of the pair, R is the coefficient of genetic relatedness, the b's are least squares regression coefficients, and e is the residual. In this model, b3 provides an estimate of heritability, h^2, and b1 provides an estimate of shared environmental variance, c^2 (DeFries & Fulker, 1985; Rodgers & McGue, 1994). Because there is no selection of one member within the kinship pair, the data are "double entered," so that each kinship pair contributes two observations, one with the first member of the pair as FM1 and the second as FM2, the other with the first member as FM2 and the second as FM1. After double entry, standard errors and significance tests may be conservatively computed by adjusting the sample size back to the number of pairs.

If either h^2 or c^2 is estimated to be around zero, that part of the model can be dropped and re-estimated to increase the power associated with the estimate of the other term. For example, if $b1=c^2=.02$ and $b3=h^2=.40$, then the following model can be estimated:

$$FM1 = b0 + b2*R + b3*(FM2*R) + e, \qquad (2)$$

which will re-estimate h^2 and give a more powerful statistical test of heritability.

The estimate of h^2 obtained from DF analysis is an estimate of the additive genetic influence, or narrow-sense heritability.   Non-additive genetic influence – associated with dominance and/or epistasis – also exist. Waller (1994) presented a modification of the DF analysis model which tests for a dominance effect if there is no significant shared environmental influence.   To estimate this model requires that dominance coefficients be defined along with the R coefficients; we defined dominance coefficients of $D=.25$ for full siblings and $D=0$ for half siblings and cousins. For our twins of unknown zygocity, we defined $D=..625$, halfway between the $D=1.0$ for MZ twins and $D=.25$ for DZ twins. For our category of unknown full or half siblings, we defined $D=.125$, halfway between full sibling's $D=.25$ and half sibling's $D=0$.   (We also re-ran the analysis with the ambiguous full/half siblings all coded $D=.25$, since the majority of those are likely to be full siblings based on base rates – results were very similar, and only the first analysis will be presented below.) Once D coefficients were defined, we fit the model suggested by Waller:

$$FM1 = b0 + b2*R + b3*(FM2*R) + b4*(FM2*D) + e. \qquad (3)$$

In this model, b3 estimates h^2, b4 estimates variance associated with dominance, and b3+b4 provides an estimate of broad-sense heritability.

One final analysis will be run that is qualitatively different from the previous analyses. Theoretical literature related to Fisher's theorem has emphasized (as Fisher did as well) that heritabilities are not appropriate for comparisons of genetic influence across domains (e.g., Burt, 1995; Houle, 1992). The reason for this is that heritabilities are defined as the proportion of phenotypic variance associated with genetic influences – but phenotypic variance will change across domains. A fixed amount of genetic influence will appear greater in relation to large phenotypic variation than in relation to smaller variation. Our theoretical framework above suggested that fertility expectations should contribute part of the variance in both phenotypic and genetic variance underlying fertility outcomes. But the heritabilities are not directly comparable. Rather, estimates of genetic variance in each trait will be estimated for this comparison.

Among several possibilities, Houle (1992) recommended use of the coefficient of variation – the standard deviation standardized by the mean of the trait – associated with additive genetic variance as a much more interpretable measure of genetic variance. To estimate the coefficient of variation associated with additive genetic variance we will use the formula

$$CV_a = 100 \sqrt{V_a} / M = 100 \sqrt{V_p} \sqrt{h^2} / M = 100 S \sqrt{h^2} / M, \qquad (4)$$

where V_a is additive genetic variance, V_p is phenotypic variance, and M and S are the trait mean and standard deviation. $S \sqrt{h^2} / M$ is obtainable from the observed trait mean, standard deviation, and estimated heritability. We will compute the CV_a values for both fertility expectations and fertility outcomes.

In an analysis of 842 analyses in the literature of traits related to reproductive fitness, Houle (1992) found that CV_a had higher values for traits more highly related to reproductive fitness. This is the reverse of what would be expected from a straightforward application of Fisher's theorem, and also the reverse of what comparisons of heritabilities show. Houle attributed low heritabilities for high fitness traits to their high residual variation, and high CV_a values to the many influences on these outcomes.

This argument has direct implications for our theoretical structure above. Individual fertility expectations are one of many influences on fertility outcomes (along with partner desires, contraception failure, infertility, and other random factors -- note that some of these can potentially also have genetic variance underlying them). As a result, we expect properly standardized measures of both phenotypic and genetic variance to be lower for fertility expectations than for fertility outcomes.

4. RESULTS

4.1   Descriptive Results

In this section, we present descriptive results from the overall NLSY dataset and from the NLSY sibling dataset. The overall dataset contained N=12686 observations in 1979, and then lost small portions of the dataset to attrition in every year after that. In 1979, the mean age was 17.9. Average ideal family size in 1979 was 2.93 (S=1.53, N=12626), average desired family size was 2.53 (S=1.53, N=12584), average expected family size was 2.36 (S=1.46, N=12460), and the average actual family size for the 1400 respondents who had one or more children was 1.30 (S=.60, N=1400). By 1982, the average ideal family size had decreased to 2.65 (S=1.15, N=12048), average desired family size was 2.40 (S=2.40, N=12032), and average expected family size was 2.34 (S=1.27, N=11989).

The expected family size can be tracked across years, since it was collected a number of times (in 1979, 1982-1986, 1988, 1990, 1992, and 1994). In Table 1, we present the pattern of means for expected family size across these years. The means show a systematic decline across years. Between 1979, when the sample ranged in age from 14-22, until 1994, when the sample ranged in age from 29-37, the expected family size declined from 2.36 to 2.13. The mean expected family size by age (instead of by years) showed the same pattern; from age 19 (N=4131) until age 34 (N=2095), mean expected family size declined systematically from 2.34 to 2.14. At the same time, the mean completed fertility necessarily increased systematically. These are shown in Table 1 also.

There is some question as to whether these decreases in expected family size were caused by changes at the individual level, or by the attrition across years. To address this question, we extracted the 7081 individual who had measures of expected family size in every one of the nine years. These results were just as systematic as those in Table 1. Mean expected family size across the nine years was 2.36, 2.34, 2.30, 2.30, 2.29, 2.22, 2.19, 2.14, and 2.14. Obviously, individuals on the average reduced their expected family size over time.

4.2   Analysis by Year

The NLSY sibling file was constructed using the kinship links described earlier. Then, kinship correlations were computed for each level of relatedness, and a DF Analysis was run to estimate h^2 and c^2. For ideal family size, $h^2 = .60$ and $c^2 = -.02$ in 1979 (N=2323 double-entered individuals) and $h^2 = .02$ and $c^2 = .13$ in 1982 (N=2249). The difference

Table 1: Descriptive Statistics (Mean and Standard
        Deviation), for Expected Fertility and
        Completed Fertility, Overall NLSY Dataset,
        by Year and by Age

Year	Mean Exp Fert	SD	N	Mean Comp Fert	SD	N
1979	2.36	1.46	12460	.14	.60	12686
1982	2.34	1.27	11989	.38	.74	12122
1983	2.30	1.50	12125	.47	.82	12220
1984	2.25	1.22	11978	.57	.90	12069
1985	2.28	1.14	10817	.67	.97	10894
1986	2.27	1.28	10584	.78	1.05	10655
1988	2.20	1.18	10347	1.04	1.18	10465
1990	2.17	1.57	10296	1.25	1.26	10436
1992	2.14	1.26	8905	1.42	1.32	9016
1994	2.13	1.30	8725	1.58	1.36	8891

Age	Mean Exp Fert	SD	N	Mean Comp Fert	SD	N
18	2.34	1.25	2590	.18	.46	2613
19	2.35	1.26	4131	.26	.56	4162
20	2.32	1.30	5594	.34	.64	5637
21	2.34	1.65	7156	.44	.75	7214
22	2.30	1.25	7463	.50	.81	7530
23	2.28	1.19	7773	.59	.90	7824
24	2.24	1.19	8812	.70	.96	8892
25	2.24	1.17	7775	.82	1.06	7842
26	2.21	1.22	7042	.91	1.08	7110
27	2.22	1.18	5754	1.04	1.18	5812
28	2.20	1.25	5341	1.13	1.19	5402
29	2.20	1.99	4388	1.25	1.27	4444
30	2.13	1.21	4971	1.32	1.27	5045
31	2.13	1.28	4445	1.46	1.34	4522
32	2.13	1.24	3520	1.51	1.30	3572
33	2.14	1.29	2989	1.61	1.37	3023
34	2.14	1.32	2095	1.69	1.37	2138

between 1979 and 1982 seemed to be caused by very similar kinship correlations for R=.375 and R=.5 in 1982 (r=.15 versus r=.14), compared to very different (and more genetically suggestive) kinship correlations in 1979 (r=.08 versus r=.26). For desired family size, h^2 = .76 and c^2 =-.20 in 1979 (N=4612) and h^2 = .46 and c^2 = -.06 in 1982 (N=2249). This pattern was more interpretable, suggesting a meaningful genetic influence on desired family size and no shared environmental influence. The small negative c^2 values indicated violations of the additive genetic model; for example, the kinship correlation for half siblings (R=.25) in 1982 was r=-.02, and for ambiguous half/full siblings (R=.375) in 1979 was r=-.08. We view these results for ideal and desired family size as suggestive of genetic influence on fertility desires. We turn to an analysis of the expected family sizes to further investigate these patterns.

In this analysis, the cohort of 14-22 year old NLSY adolescents in 1979 is followed from 1982 until 1994, accounting for their expected and actual family sizes at regular intervals (note that the sibling sample sizes were too small to do this analysis in 1979). In each year, we did a DF Analysis to estimate h^2 and c^2. These results are shown in Table 2. Results are consistent across years, suggesting a moderate heritability and no shared environmental influence. The median h^2 and c^2 estimates across these nine separate analysis was h^2 = .42 and c^2 = -.06. These results are obviously non-significant in the context of the large standard errors, but their consistency is highly suggestive of moderate genetic influence. Further, when the model was re-estimated using equation 2, the heritabilities were highly significant. This procedure will be illustrated in more detail in the analysis by age below. (Note that all standard errors in this paper have been adjusted from those estimated by our procedures by multiplication by $\sqrt{2}$, which is a conservative adjustment that accounts for the double-entry procedure. As a result, standard errors should be considered to be conservative upper bound estimates of the actual standard errors.)

The median kinship correlations underlying these analyses were r=.08 (N=142 double entered individuals) for R=.125 (cousins), r=.00 (N=84) for R=.25 (half siblings), r=.05 (N=586) for R=.375 (ambiguous half/full siblings), r=.16 (N=3656) for R=.50 (full siblings), and r=.14 (N=56) for R=.75 (same-sex twins of unknown zygocity). There are minor violations of the additive genetic model among these kinship correlations (leading to negative c^2 values) but the general pattern is clearly consistent with genetic influences on fertility expectations.

We also ran the equivalent analysis on completed fertility, for which we had measures across the same time period for which we had information about fertility expectations. Results of the DF Analysis are contained in

Table 2: Genetic Variance (h²) and Shared Environmental Variance (c²), NLSY Kinship Cohort Sample, by Year

Year	Fertility Expectations					Fertility Outcomes				
	h^2	se-h^2	c^2	se-c^2	N	h^2	se-h^2	c^2	se-c^2	N
1979	***	***	***	***	88	***	***	***	***	88
1982	.27	.26	.08	.13	4450	.73	.19	-.05	.09	4544
1983	.29	.27	.13	.13	4514	.51	.19	.05	.09	4584
1984	.42	.26	-.04	.12	4502	.43	.20	.08	.10	4552
1985	.64	.26	-.15	.12	4426	.09	.20	.25	.10	4492
1986	.60	.26	-.14	.11	4378	.01	.20	.25	.10	4426
1988	.54	.23	-.15	.11	4516	-.12	.22	.29	.11	4634
1990	.39	.26	-.06	.12	4200	.04	.22	.19	.11	4328
1992	.12	.68	.05	.74	3690	.04	.24	.16	.11	3800
1994	.54	.26	-.18	.12	3532	.03	.25	.14	.12	3674
Median	.42	---	-.06	---		.04	---	.16	---	

Table 2 along with those for fertility expectations. The median heritability and shared environmental variance estimates for the nine years were $h^2 = .04$ and $c^2 = .16$, both non-significant. There was a fairly substantial range for each of these (h^2 ranged from -.12 to .73, and c^2 ranged from -.05 to .29). These medians suggest no genetic influence on completed fertility, and possible small shared environmental influence.

Closer inspection of these results suggests a slightly different interpretation, however. The h^2 values systematically decreased across year, eventually dropping to zero. They are, in fact, of moderate to large size for the first three years. Since the completed fertility up to 1984 and before applies to the cohort when it was relatively young (age 20-28 in 1985), this may imply different patterns at different ages. We will further inspect this possibility when we do a set of DF Analyses by age in the next section.

4.3   Analysis by Age

In addition to the analysis in which the cohort of NLSY respondents were observed across years, we also constructed a different data structure to investigate age patterns. In this analysis, we matched NSLY kin at a fixed age by using the time structure of the NSLY fertility expectation responses. In this analysis we have the advantage of not having to adjust for age differences statistically. While the sample sizes are somewhat smaller in our age analysis, we view this as more than compensated by the ability to age match within our kinship pairs. We report ages for which we had 1000 or more respondents from 500 or more kinship pairs, which gave us h^2 and c^2 estimates for nine ages from 20 to 28. These results are presented in Table 3. The median estimate for heritability and shared environmental influence for expected fertility across these nine ages was $h^2 = .60$ and $c^2 = -.14$. Standard errors were large enough that even large heritability estimates were non-significant (but re-fitting equation 2 results in significant heritabilities, as we will show). The kinship correlations underlying these estimates had a median r=.11 for R=.125 (cousins, median N=112), median r=.32 for R=.25 (half siblings, median N=80), median r=.20 for R=.375 (ambiguous half/full siblings, median N=288), median r=.23 for R=.50 (full siblings, median N=1780), and median r=.36 for R=.75 (same-sex twins of unknown zygocity, median N=36). These patterns are supportive of moderate and meaningful genetic influence and no shared environmental influence. The half-sibling pairs have kinship correlations that are higher than would be predicted by an additive genetic model, otherwise results are consistent with an additive genetic model.

Following the analysis of fertility expectations by age, we ran a set of DF Analyses of completed fertility by age. Completed fertility for those

Table 3: Genetic Variance (h^2) and Shared Environmental Variance (c^2), with Standard Errors (se's), NLSY Kinship Age Sample, by Age

Age	Fertility Expectations					Fertility Outcomes				
	h^2	se-h^2	c^2	se-c^2	N	h^2	se-h^2	c^2	se-c^2	N
20	.63	.42	-.21	.20	1456	.33	.36	.03	.17	1492
21	1.21	.48	-.48	.19	2302	.44	.25	-.01	.11	2348
22	-.07	.39	.13	.19	2018	.07	.29	.16	.13	2064
23	.44	.41	-.10	.19	1996	-.03	.29	.23	.13	2012
24	.60	.33	-.19	.15	2386	-.03	.28	.32	.13	2436
25	.29	.44	.02	.21	1892	.40	.29	.17	.14	1930
26	.50	.38	-.10	.18	1622	.45	.36	.10	.17	1646
27	.73	.54	-.21	.26	1236	.41	.38	.10	.18	1262
28	.64	.57	-.14	.27	1062	.17	.48	.16	.23	1080
Median	.60		-.14			.33		.16		

aged 20 should have a different meaning than for those in their late 20's, since the early ages account primarily for those who had children in their teens. But it is important to note that, for all ages, these are substantially truncated fertility outcomes compared to completed fertility. At ages 20 and 21 h^2 was small, then it became essentially zero for ages 22-24, and then jumped up to a moderate level for ages 25-28. The median values across the nine ages were $h^2 = .33$ and $c^2 = .16$ (neither significant), suggesting the possibility of genetic and shared environmental influences on completed fertility defined by age.

Because we were more comfortable with the age adjustment procedure from the analysis by age, we re-fit the DF model to the dataset by age in several additional analyses. The first was a test for dominance, which involved re-fitting the model using equation 3. A pattern suggestive of possible dominance effects is one that estimates negative c^2 values, which was a fairly typical finding for the fertility expectations in Table 3 (which had median $c^2 = -.14$). However, eight of the nine estimates for dominance (the b4 values in Equation 3) were negative, and none were significant, a finding which was inconsistent with dominance effects.

If there are no dominance effects, it is entirely possible that the negative c^2 values in Table 3 for fertility expectations were caused by a zero value in the population and some sampling variability. In this case, re-estimating the model by dropping the estimate of the c^2 provides a more powerful statistical test of the heritabilities. In Table 4 are listed the fertility expectation heritabilities and their standard errors obtained from fitting equation 2, assuming no shared environmental influence. The median h^2 was .24, and this estimate was highly significant. It is worth noting that the median h^2 in Table 3 of .60 was non-significant, while the much lower median h^2 of .24 in Table 4 was highly significant – we have consistently found the model estimating only a genetic or only a shared environmental component to deliver substantially increased power.

4.4   Analysis of Genetic Variance Estimates

A final analysis is one that enlarges the scope of interest to include estimates of genetic variance. Theoretical work related to Fisher's fundamental theory has criticized the use of heritabilities for comparative purposes, as discussed in the analysis section above. In Table 5 we present CV_a values, which are standardized estimates of genetic variance that are comparable across domains (as heritabilities are not). These were obtained by applying equation 4 to the heritabilities in Table 3 and 4, and using the means and standard deviations from the overall NLSY dataset presented in Table 1 as phenotypic measures. It is not obvious which set of heritabilities

Table 4: Genetic Variance-only (h^2) model for Fertility Expectations, NLSY Kinship Sample, by Age

Age	Fertility Expectations		
	h^2	$se\text{-}h^2$	N
20	.19	.08	1456
21	.04	.07	2302
22	.20	.06	2018
23	.24	.07	1996
24	.22	.06	2386
25	.32	.07	1892
26	.30	.07	1622
27	.30	.08	1236
28	.35	.09	1062

Median	.24		

Table 5: Heritabilities and Coefficients of Variation, NLSY Kinship Sample, by Age

Age	Fertility Expectations				Fertility Outcomes	
	Table 3		Table 4		Table 3	
	h^2	CV_a	h^2	CV_a	h^2	CV_a
20	.63	44.5	.19	24.4	.33	124.9
21	1.21	77.6	.04	14.1	.44	113.1
22	-.07	(0)	.20	24.3	.07	42.9
23	.44	34.6	.24	25.6	-.03	(0)
24	.60	41.2	.22	24.9	-.03	(0)
25	.29	28.1	.32	29.5	.40	81.7
26	.50	39.0	.30	30.2	.45	79.6
27	.73	45.4	.30	29.1	.41	72.7
28	.64	45.5	.35	33.6	.17	43.4
		-----		-----		-----
Median		41.2		25.6		72.7

gives the best estimate of genetic variance accounted for in the fertility expectations measures, and that is why results from both Table 3 and 4 are presented. The interpretation is identical in each case.

Consistently, the fertility outcomes show higher levels of standardized genetic variance than the fertility expectations. The median CV_a for the fertility expectations were 41.2 and 25.6 for Table 3 and Table 4 heritabilities, respectively. The median CV_a for the fertility outcomes was 72.7, approximately twice the magnitude of the median CV_a's for fertility expectations.

5. DISCUSSION

These results support our prediction that fertility expectations/desires would have a moderate heritability, and fertility outcomes would have a smaller heritability. Our prediction was derived both from our theoretical framework and from past empirical research. The evidence for genetic influences on fertility expectations was consistent and compelling, as was that for fertility desires (and, to a lesser extent, fertility ideals). The evidence for genetic influence on fertility outcome was more tentative, and seemed to differ across both year and age. Nevertheless, results were suggestive of small genetic influences on completed fertility as well.

These findings appear to be consistent with the proposal in our theoretical framework. It may well be that fertility expectations/desires must be influenced genetically in a fairly robust and consistent way before they can have any influence on fertility outcomes, since they are competing for influence alongside many random and environmental factors. Thus, the amount of heritability in the fertility outcomes may well indicate the direct influence of the fertility expectations/desires, attenuated by these other factors.

We found no evidence of shared environment influences on fertility expectations. Our results were suggestive that the additive genetic model was violated in some minor ways, although these violations were fairly minor. When we tested for dominance effects -- one possible cause of such deviations -- we found no evidence of this type of non-additive genetic influence. We did find some suggestive evidence that there may be small shared environmental influences on fertility outcomes, although these apparently are not translated through fertility desires and expectations. We can speculate that these may be responses from childbearing parents to suggestions from parents and friends that are, in fact, inconsistent with the desires/expectations of the individual themselves. For example, a couple might well wish their childbearing complete at one child, but may also be influenced by suggestions from the grandparents that more children is better, or that the child needs a sibling. This type of influence is consistent with shared environmental influence on fertility outcomes that would not show up in fertility desires/expectations.

At the same time that we found a higher proportion of genetic influence on the variance to be explained in fertility expectations than in fertility outcomes, we also did an analysis that accounted for the overall variance alongside the heritabilities. Again, results were in the direction predicted by previous theory and empirical results. Fertility expectations are one of a number of possible influences on fertility outcomes, alongside contraceptive failures, infertility, partner desires, and random processes. Each of these – especially fertility expectations – are conceptually prior, and therefore causes of fertility outcomes, and not the other way around. Thus, we expect the amount of both phenotypic and genetic variance to be greater for fertility outcomes than for any of its component parts. This expectations was supported.

Thus, to summarize, there was greater genetic variance in fertility outcomes than fertility expectations, suggesting the potential for other influences besides fertility expectations to contribute to genetic variance in fertility outcomes. But the genetic variance that was there was a smaller proportion of the total phenotypic variance for fertility outcomes than for fertility expectations (reflected in the lower heritabilities for outcomes than for expectations). This suggests a sensible and clean framework. Fertility expectations are genetically influenced, and around half of their overall variation is caused by genetic influences. Shared environmental influences play no role, so that the other half is attributable to nonshared environmental influences and measurement error. At least part of the genetic variance in expectations becomes variance in fertility outcomes, but there are other sources of genetic influences on fertility outcomes as well, as indicated by the higher levels of genetic variance shown in Table 5 for outcomes than for expectations.

An analysis that we attempted, but which did not lead to interpretable results, was a linking analysis in which we tested for genetic influences in measures of the link from expectations to outcomes. There are difficult methodological questions involved in this linking process, including in particular the issue of scaling the different measures into a common metric. We will be working on this methodology, and we hope to present results of such an analysis in future research.

We conclude with the usual call for more research, but with the ability to provide more direction for that research than we had in this project. As Miller and his colleagues predicted, we found empirical support for the existence of genetic influences on fertility desires/expectations. Further, we embedded those findings in a bigger framework linking fertility expectations/desires to fertility outcomes. And small genetic influences appeared likely on fertility outcomes as well. The findings need to be replicated, but there are clear suggestions that should guide predictions for those replications.

REFERENCES

Burt, A. (1995). The evolution of fitness. *Evolution, 49*, 1-8.

DeFries, J. C. & Fulker, D. W. (1985). Multiple regression analysis of twin data. *Behavior Genetics, 15*, 467-473.

Dunne, M. P., Martin, N. G., Statham, D. J., Slutske, W. S., Dinwiddie, S. H., Bucholz, K. K., Madden, P. A., & Heath, A. C. (1997). Genetic and environmental contributions to variance in age at first sexual intercourse. *Psychological Science, 8*, 1-6.

Edwards, A. W. F. (1994). The fundamental theorem of natural selection. *Biological Review, 69*, 443-474.

Ewens, W. J. (1989). An interpretation and proof of the fundamental theorem of natural selection. *Theoretical Population Biology, 36*, 167-180.

Fisher, R. A. (1930). The genetical theory of natural selection. New York: Dover Publishers.

Houle, D. (1992). Comparing evolvability and variability of quantitative traits. *Genetics, 130*, 195-204.

Lessard, S. (1997). Fisher's fundamental theory of natural selection revisited. *Theoretical Population Biology, 52*, 119-136.

Martin, N. G., Eaves, L. J., & Eysenck, H. J. (1977). Genetical, environmental, and personality factors influencing the age of first sexual intercourse in twins. *Journal of Biosocial Science, 9*, 91-97.

McGue, M. & Lykken, D. T. (1992). Genetic influences on risk of divorce. *Psychological Science, 3*, 368-373.

Mealey, L. & Segal, N. L. (1993). Heritable and environmental variables affect reproduction-related behaviors, but not ultimate reproductive success. *Personality and Individual Differences, 14*, 783-794.

Miller, W. B., Comings, D. E., MacMurray, J., Chiu, C., Wu, H., & Pasta, D. J. (1999a). Dopamine receptors are associated with age at first sexual intercourse. *Journal of Biosocial Sciences, 31*, 43-54.

Miller, W. B., Pasta, D., MacMurray, J., Chiu, Wu, Comings, D. (1999b). Genetic influences on childbearing motivation and parental satisfaction: A theoretical framework and some empirical evidence. In L. Severy & W. Miller (Eds) *Advances in Population: Psychosocial Perspectives, vol. 3*. Jessica Kingsley.

Perusse, D., Neale, M. C., Heath, A. C., & Eaves, L. J. (1994). Human parental behavior: Evidence for genetic influence and potential implication for gene-culture transmission. *Behavior Genetics, 24,* 327-335.

Plomin, R., DeFries, J. C., & McClearn, G. E. (1990). *Behavioral genetics: A primer.* New York: Freeman.

Rodgers, J. L. (1996) NLSY Linking Algorithm. Unpublished document.

Rodgers, J. L. & McGue, M. (1994). A simple algebraic demonstration of the validity of DeFries-Fulker analysis in unselected samples with multiple kinship levels. *Behavior Genetics, 24,* 259-262.

Rodgers, J. L., Rowe, D. C., & Buster, M. (1999). Nature, nurture, and first intercourse: Fitting behavior genetic models to NLSY kinship data. *Journal of Biosocial Sciences, 31,* 29-41.

Ryder, N. M. & Westoff, C. F. (1969). Relationship among intended, expected, desired, and ideal family size: United States, 1965. *Population Research* (NICHD).

Williams, L. A. & Williams, B. J. (1974). A re-examination of the heritability of fertility in the British peerage. *Social Biology, 21,* 225-231.

Chapter 5

A HERITABILITY STUDY OF CHILDBEARING MOTIVATION

David J. Pasta and Warren B. Miller

Key words: childbearing motivation, childbearing questionnaire (CBQ), positive childbearing motivations (PCM), negative childbearing motivations (NCM), DF analysis, heritability

Abstract: Childbearing motivation can be thought of as a latent disposition to feel motivated for or against childbearing. Previous work has explored the antecedents of childbearing motivation and placed the trait of childbearing motivation in a conceptual sequence with desires, intentions, and behaviors. This study assessed the heritability of childbearing motivation as measured by the two main scales of the Childbearing Questionnaire (CBQ) and their nine subscales.

We used DF analysis (DeFries and Fulker, 1985) to assess heritability of childbearing motivation in a sample of married couples living in Santa Clara County, California and their relatives. Respondents were administered the Childbearing Questionnaire five times over a five year period. Relatives were asked to participate in a brief telephone interview and complete the CBQ. We analyzed the 278 respondents who lived with their relative(s) during childhood and their 399 relatives who completed the CBQ. We also included 174 relative-relative pairs who had lived together during childhood. Of the 573 pairs, 539 (94%) are non-identical twins or full siblings with average genetic relatedness R=0.5.

We found modest to strong evidence of heritability for both the main scales and all but one subscale of the CBQ. Despite the limitations of the sample, the results were robust to sensitivity analysis.

1. BACKGROUND

For the past ten years, we have been studying adult human reproductive behavior in a sample of 401 married couples living in Santa Clara County, California (Miller, 1992). In November 1988, we drew a modified area probability sample in which, by design, 203 couples had no children and 202 couples had one child. The mean age of the wives was 29.6 and of the husbands was 31.6. The obtained sample was predominantly middle class: mean years of education and mean income (in $1000s) were 15.9 and 25.7 for the wives and 16.7 and 45.9 for the husbands. Non-Hispanic European-Americans (i.e., white) made up 81% of the sample, with the remainder being African-American, Asian-American, or Latin-American. Each respondent was initially interviewed in person and then administered a battery of self-report inventories. Four follow-up interviews were conducted at one, two, three and one half, and five years after the initial interview. At each of the five interviews, respondents were asked about their child-number desires (Miller and Pasta, 1993) and their childbearing desires and intentions (Miller, 1994), as well as being administered the Childbearing Questionnaire (Miller, 1995), a measure of childbearing motivation.

Childbearing motivation may be thought of as a motivational trait, that is, as a latent disposition to feel motivated for or against childbearing. The Childbearing Questionnaire (CBQ) was designed to measure both the positive and negative components of childbearing motivation. The two main scales of the CBQ, which are essentially uncorrelated with each other, are Positive Childbearing Motivation (PCM), based on 27 items, and Negative Childbearing Motivation (NCM), based on 20 items. Each item describes a behavior, feeling, or situation typically associated (in the contemporary United States, at least) with having a child. The respondent is asked to indicate the desirability (for PCM) or undesirability (for NCM) of each item on a four-point scale: Very, Moderately, Slightly, and Not. The male and female versions of the questionnaire are worded slightly differently as appropriate. PCM has five subscales and NCM has four subscales (for additional information, see Miller et al., 1999, in this volume).

Previous work has led to a theoretical framework for understanding childbearing motivation and its possible heritability. Miller (1992) examined the antecedents of childbearing motivation. These antecedents can be divided into those that are more experientially based, such as the quality of the mother's parenting; those that are primarily inherited, such as personality traits such as Nurturance or Affiliation (Jackson, 1984); and those that more likely represent an interaction between genes and experience, such as enjoyment of child caretaking as a teenager. Interestingly, there are significant sex differences in both the experience-based and the inherited antecedents.

Earlier work has also placed the trait of childbearing motivation in a conceptual sequence with desires, intentions, and behaviors (Miller, 1994; Miller & Pasta, 1993; 1994; 1995). Desires represent psychological states that express how an individual feels about specific goals. They derive from motivations but are also influenced by attitudes, beliefs, and other traitlike factors. Desires are to some degree stable over time, but less so than traits. With respect to the domain of fertility behavior, there are three important desires. Childbearing desire is the desire for a child, child-number desire is the desire for a certain number of children, and child-timing desire is the desire to have a child at a particular time. Intentions represent psychological states that express what an individual actually plans to do. Intentions are largely derived from desires but are shaped by reality constraints and, in particular, by partner desires. There are three types of intentions corresponding to the three types of desires. Behavior is what the individual actually does in order to accomplish his or her goals. In the fertility domain there are two primary types of such instrumental behavior: proceptive behavior (efforts to achieve conception; Miller, 1986) and contraceptive behavior. The earlier studies of the traits, desires, intentions, behavior sequence revealed a number of sex differences.

1.1 Estimating Heritability

To assess heritability, behavioral geneticists have historically relied heavily upon twin studies. In twin studies, the correlation between cotwins on the outcome measure is compared for "identical" (monozygotic, or MZ) twins and for fraternal (dizygotic, or DZ) twins. Under the assumption that cotwins have identical environmental influences, any systematic difference between the correlations for MZ and DZ twins must be due to genetic effects. The genetic influence can be quantified using the fact that MZ twins share all their genes but DZ twins share on average only half of their genes. This quantification gives rise to the formal definition of heritability as the proportion of observed variability in an outcome measure that is attributable to genetics. The residual variability – that not explained by genetics – is by definition ascribed to environment and has been called "environmentality" (Fuller and Thompson, 1978). By its definition as the residual variability, environmentality includes any interaction between genetics and environment.

In twin studies and the extensions discussed here, authors generally assume negligible assortative mating and an additive genetic model. Equal shared environments across kinship categories are also usually assumed, although this assumption is questionable even for twins. Although twin studies are very useful tools for behavioral genetics, they are often difficult to do in practice. Only about 1 in 83 births results in twins, so the vast majority of the population is necessarily ineligible for twin studies. In recent years, increasing attention has turned to the study of individuals with more general genetic relationships.

Although in classic twin studies it is not difficult to calculate estimates of heritability directly from the correlations within the MZ and DZ subgroups, with more complicated studies it is much more convenient to use statistical estimation techniques implemented on a computer to fit a quantitative genetic model. Most commonly, linear structural equations software such as LISREL (Jöreskog and Sörbom, 1996) or special-purpose software is used to calculate maximum-likelihood estimates of the parameters. This approach has many advantages but has the disadvantage of being somewhat cumbersome to apply, especially at the exploratory stage of analysis.

An alternative to maximum-likelihood estimation of simultaneous equations employs ordinary least squares multiple regression. DeFries and Fulker (1985) described this approach in the context of twin data with a trait-selected sample (where one member of the twin pair has been identified because of an unusual characteristic). The form of analysis they introduced, referred to as DF analysis, has been extended to nontwins and to unselected samples by various authors.

In the original formulation of DF analysis, one member of the twin pair (the proband) was trait-selected. The basic regression model expressed the score for the cotwin (C) as a linear function of the score for the proband (P) and a variable R that codes the genetic relatedness of the twins, with R taking the value 1.0 for MZ twins and 0.5 for DZ twins:

$$C = B_1 * P + B_2 * R + A \qquad (1)$$

Adding an interaction term, the product of the proband's score P and the genetic relatedness R, produces an "augmented" regression model:

$$C = B_3 * P + B_4 * R + B_5 * P * R + A \qquad (2)$$

For trait-selected samples, this model permits a direct estimate of heritability (B_5, the coefficient of the interaction term), a direct estimate of the shared environmental effect (B_3, the coefficient of the proband's score), and a direct test of the assumption of equal shared environmental influences for the two types of twins (B_4, the coefficient of R, which should be zero if the assumption is met).

This model is sometimes referred to as the ACE model, referring to the additive genetic variance (A), the common environmental variance (C), and the nonshared environmental variance (E). In practice, the common environmental variance, c^2, may be estimated as a negative value. This is an indication that the assumed model does not adequately fit the data. If the coefficient B_3 of the proband's score P is not statistically significant, it is common practice to drop the corresponding term from the regression model and estimate the AE model. This is potentially somewhat misleading. When the true model includes a common environmental effect, omitting the corresponding term is a misspecification of the model and leads to biased estimates of the other parameters and in particular a biased estimate of the additive genetic variance, h^2. Omitting that term from the regression model corresponds to omitting nonsignificant main effects in an analysis of variance but retaining an interaction term. Although this simplifies the model and produces smaller estimated standard errors for the remaining coefficients, it can be capitalizing on chance or a small sample size. The standard errors are underestimated if the true model includes some common environmental effect, no matter how small. In practical applications of DF analysis, there may be a high correlation between the proband's score P and the interaction of genetic relatedness and the proband's score P*R. A high correlation between these two variables leads to high standard errors of the corresponding coefficients to reflect the near collinearity. Removing one of the terms would then lead to much lower estimated standard errors but this may be at least partly artificial. As long as the resulting statistical tests are treated with some caution, however, estimating the heritability h^2 from the AE model certainly provides valuable additional information.

When the estimate of c^2 from an ACE model is negative and statistically significant, it is of questionable legitimacy to estimate an AE model. One alternative is to fit a model that omits any shared environmental effect and instead estimates an additive genetic variance and a dominance genetic variance along with a nonshared environmental variance. This ADE model, developed by Waller (1994), is estimated using the equation

$$C = B_4*R + B_5*P*R + B_6*P*D + A \qquad (3)$$

where R is as usual the degree of additive genetic relatedness (1.0 for MZ, 0.5 for DZ) and D is the coefficient of dominance genetic relatedness (1.0 for MZ, 0.25 for DZ). As usual, B_5 is an estimate of the additive genetic variance h^2, B_6 is an estimate of the dominance genetic variance, d^2, and the sum of the two coefficients is an estimate of the total genetic variance.

The extension of DF analysis to unselected samples, to data that include more than two relationships, and to include other variables as predictors of environmental influences, are all mentioned in the original description (DeFries and Fulker, 1985). The details of the extensions and their application have been made in a number of publications in recent years. LaBuda et al. (1986) provided expected partial regression coefficients from models (1) and (2). In addition, they formulated a model that includes both "affected" (trait-selected) and control (unselected) twin pairs. They employ dummy variable coding designed to estimate a pooled effect and a differential effect for the affected and control groups.

Zieleniewski et al. (1987) extended DF analysis to include twins and siblings in a trait-selected sample. They include dummy variables to index separate environmental effects for twin pairs and for sibling pairs. This follows the suggestion in DeFries and Fulker (1985, p. 472) for the analysis of data from more than two relationships simultaneously: "However, it would be necessary to include additional coefficients as dummy variables produced by probands' scores to model shared environmental influences that vary as a function of relationships." The problem with this approach is that the estimate of heritability relies entirely on the twins and makes no use of the data from other kinship categories at all.

One way around this problem is to assume that all the kinship categories have equal shared environment. This assumption may be reasonable for twins and full siblings, but it is questionable for half siblings and is implausible for more distant relationships. Instead of assuming equal shared environment for each kinship category, one could model the predictors of shared environment from measured characteristics of the pair of relatives.

Our approach in this study is to include only those pairs who lived together during their childhood. This corresponds to the assumption that the shared environment is equal across kinship categories if the individuals lived together at some point during their childhood.

In unselected samples, the choice of proband and cotwin is arbitrary. To account for this, Detterman et al. (1990) did ten analyses for ten alternative random designations of one member of each pair to be the proband. This is an approximation to analyzing all 2^N possible designations of one member as the proband and one as the cotwin. A simpler approach to the problem of which twin to consider as the proband is to "double enter" the data, treating each member of the twin pair as proband in one observation and as cotwin in another observation. This approach has a long history (Haggard, 1958) and was used in one of the first applications of DF analysis to unselected data (Rodgers and Rowe, 1987). Double entry produces unbiased estimates (Cherny, DeFries, Fulker, 1992, p. 493). However, its use does necessitate adjustment of the standard errors and degrees of freedom to reflect the duplication of data. Cherny, Cardon, Fulker and DeFries (1992, p. 158) found parameter estimates obtained from the average of ten randomized trials and the double entry method to be "highly similar."

2.   METHODS AND RESULTS

2.1   Data Used in Heritability Study

At the conclusion of the longitudinal study of husbands' and wives' childbearing motivation, we asked respondents for permission to contact relatives of theirs so that we could briefly interview them by telephone and ask them to complete the Childbearing Questionnaire (CBQ).   Loss to follow-up was primarily due to separation and divorce.

Table 1.   Relationship and genetic relatedness (R) of respondent-relative, relative-relative, and total pairs.

Relationship	Genetic Relatedness (R)	Number of Resp-Rel Pairs	Number of Rel-Rel Pairs	Total Pairs
Identical Twins	1.0	6	2	8
Non-identical Twins/ Full Siblings	0.5	376	163	539
Half Siblings	0.25	4	1	5
First Cousins	0.125	1	3	4
Step Siblings/ Adopted Siblings	0.0	12	5	17
		399	174	573

By the fourth and final follow-up interview, 694 (87%) of the participants remained in the study.   Of the 694 in the study at that time, 338 (49%) provided a relative who participated.   We consider here only the 278 respondents who lived with their relative(s) during childhood and their 399 relatives who completed the CBQ.   In addition to the 399 respondent-relative pairs, we also included 174 relative-relative pairs who had lived together during childhood.   One relative was provided by 183 respondents, two relatives by 76 respondents, three relatives by 13 respondents, four relatives by 5 respondents, and five relatives by 1 respondent.   Table 1 gives the relationship and genetic relatedness (R) of the respondent-relative, relative-relative, and total pairs.

In this study, we had multiple CBQ scores for most of our respondents (five from 374, four from 13, three from 6, and one from 6).   Selecting a single CBQ score from the respondent would not have made full use of the available data. Averaging the multiple CBQ scores for the respondents would have distorted the data by making the information for respondents appear less variable than that for their relatives. Accordingly, we compared each respondent with his/her relative(s) once for each respondent CBQ score. However, in order to prevent the analysis from emphasizing respondent-relative pairs over relative-relative pairs, the scores of each respondent-relative pair was down-weighted by the number of CBQ scores available for each respondent.

This "cloning" procedure has the advantage of increasing the power of our analyses. However, it also runs the risk of producing an estimated standard error in the statistical tests that is somewhat too low. We found that the standard deviation across interviews within respondents was about 30% less than the standard deviation across members of the same family. However, we would expect the standard deviation to be smaller even if we had multiple CBQ scores from different individuals with the same genes (i.e., MZ twins) and not the same person at each interview. In the absence of an appreciable number of MZ twins in our study, it is impossible for us to calculate how much of a reduction in variability would be observed in MZ twin pairs compared with siblings. However, that reduction probably would be substantial, quite possibly approaching the 30% reduction in standard deviation obtained within respondents. Therefore, we believe that the significance tests reported here are at most only slightly affected by this "cloning" procedure.

The presence of multiple assessments for our respondents permitted us to perform an additional analysis as though those responses were from monozygotic twins. Because the variability in response from the same person at different times is less than one would expect between monozygotic twins, this additional analysis produces an overestimate of the heritability. If that estimated heritability were found to be low or moderate, it would provide a useful estimated upper bound on the true heritability. The specific approach taken in this analysis was to form all possible pairs of responses from the respondent, in both orders (i.e., double-entered), and treat them as having a genetic relatedness (R) of 1.0. Thus a respondent with five responses would have ten pairs formed and then double-entered to become a total of 20 pairs.

2.2 Estimates of Heritability

We conducted DF analyses using the ACE model for the two main scales of the Childbearing Questionnaire (CBQ), Positive Childbearing Motivation (PCM) and Negative Childbearing Motivation (NCM), as well as five PCM and four NCM subscales. In Table 2 we show the estimated heritability coefficients, h^2, their associated t values, and the two-tailed probabilities (p). The estimated heritabilities are obtained as the regression parameter estimates of the interaction between the CBQ score and the genetic relatedness, R.

Table 2. Heritability Coefficients (h^2), t-values, and p-values for PCM, NCM, and nine subscales.

Scales	h^2	t	p
Subscales			
1. Positive Childbearing Motivation (PCM)	.270	1.321	.19
a. Joys of Pregnancy, Birth and Infancy	.308	1.309	.19
b. Traditional Parenthood	.562	3.029	.002
c. Satisfactions of Childrearing	.490	1.883	.060
d. Feeling Needed and Connected	.226	1.245	.21
e. Instrumental Values of Children	.034	0.180	.86
2. Negative Childbearing Motivation (NCM)	.406	2.082	.037
a. Discomforts of Pregnancy and Childbirth	.329	1.693	.090
b. Fears and Worries of Parenthood	-.032	-0.162	.87
c. Negatives of Childcare	.366	1.725	.085
d. Parental Stress	.724	3.930	<.001

The results indicate the presence of heritability for NCM, the NCM subscale Parental Stress, and the PCM subscale Traditional Parenthood. Overall, 6 of the 11 tests show statistical significance at the 0.10 level. In addition, three other heritability coefficients are at least 0.2 (PCM; Joys of Pregnancy, Birth, and Infancy; and Feeling Needed and Connected), suggesting a trend that might achieve statistical significance in other analyses, especially those with a large sample.

The estimates of the shared environmental effect, c^2, from the ACE model are shown in Table 3. Only three of the estimates are positive and the remaining eight are negative. However, the estimated c^2 is statistically significantly different from zero even at the 0.10 level for only two of the measures: Negative Childbearing Motivation (p=0.100) and Parental Stress (p=0.002). A significantly negative estimated c^2 is an indication that the assumed model does not provide an adequate fit to the data. Accordingly, these two measures were refit using the dominance model (ADE). This resulted in reasonable estimates of additive and dominance genetic variance and a sum that is statistically significantly greater than zero, indicating the presence of heritability (Table 3).

For the other nine measures, where the estimated c^2 from the ACE model was not significantly different from zero, heritabilities were estimated from the AE model (Table 3). The AE model produced much smaller estimated standard errors, leading to statistically significantly positive heritabilities for ten of the eleven measures. As discussed above, the reduction in estimated standard errors may be due in large part to the elimination of one of two variables that are highly intercorrelated. We found the correlations between P and P*R ranged from 0.63 to 0.88 over the measures. Thus the statistical tests from the AE model should be viewed with some caution. Nonetheless, the heritabilities from the ADE and AE models summarized in Table 3 provide evidence of heritability for ten of the eleven measures considered; only Fears and Worries of Parenthood shows no evidence of heritability on these data[1].

2.3 Sensitivity Analyses

As seen in Table 1, the vast majority of pairs are non-identical twins or full siblings (539 of 573, or 94%). Thus the estimates of heritability rest on the anchor of the results for pairs with R=0.5 and the results for just 34 pairs with other values of R. Under the assumptions of the model, the data for the pairs with R=0.5 lie in a straight line with the data for the other values of R. We were concerned that the estimated straight line could be substantially distorted by the presence of such a large number of pairs with R=0.5. Accordingly, as a test of the assumptions underlying the model and the sensitivity of the model to the presence of the pairs with R=0.5, we repeated the DF analysis using only the 34 pairs with other values of R.

One expected effect of elimination of 94% of the data was to increase the estimated standard errors. The increase in estimated standard error ranged from 8% to 35% with a median increase of 28%. That these increases were so modest supports the use of the original standard errors as appropriate even in the presence of such a large design imbalance.

The other expected effect of reducing the sample to the 34 pairs with values of R other than 0.5 was to change the estimated heritabilities. We anticipated that the heritabilities estimated with the reduced sample would generally be within about two standard errors of the original values. We found this was the case for eight of the eleven measures. The exceptions were PCM and two of its subscales (Joys of Pregnancy, Birth, and Infancy,

Table 3. Shared Environment Coefficients (c²), Heritability Coefficients (h² and d²), t-values, and p-values for PCM, NCM, and nine subscales.

Scales / Subscales	c^2	ADE Model					AE Model		
		h^2	d^2	h^2+d^2	t	p	h^2	t	p
1. Positive Childbearing Motivation (PCM)	.042						.353	8.219	<.001
a. Joys of Pregnancy, Birth and Infancy	.020						.348	8.082	<.001
b. Traditional Parenthood	-.093						.380	8.879	<.001
c. Satisfactions of Childrearing	-.156						.180	4.127	<.001
d. Feeling Needed and Connected	-.050						.128	2.944	.003
e. Instrumental Values of Children	.150						.327	7.590	<.001
2. Negative Childbearing Motivation (NCM)	-.160	.028	.121	.149	2.559	.010			
a. Discomforts of Pregnancy and Childbirth	-.103						.129	2.968	.003
b. Fears and Worries of Parenthood	-.003						-.037	-0.851	.39
c. Negatives of Childcare	-.081						.208	4.794	<.001
d. Parental Stress	-.291	.127	.065	.192	2.685	.007			

and Instrumental Values of Children), for which the estimated heritability was reduced by more than two standard errors to a negative value. None of those three measures had a statistically significant heritability in the original analysis. The other three subscales of PCM had estimated heritabilities that went down but by much less than two standard errors. For NCM and its four subscales, the estimated heritabilities all went up by less than two standard errors. Although it is intriguing that the estimated heritabilities increased for NCM and its subscales but decreased for PCM and its subscales, this may be due to the specifics of the 34 pairs contributing to this sensitivity analysis.

We believe that the removal of the pairs with R=0.5 provided the most valuable sensitivity analysis for our sample because of the large fraction of pairs with that value. In other studies, using the bootstrap (Efron, 1981) might lead to a better understanding of the sensitivity of the heritability estimates. Using the bootstrap in our study would not have provided a direct assessment of the extent to which the assumption of linearity was being violated by the R=0.5 pairs.

In another sensitivity analysis, we created additional pairs by treating the pairs of responses from the same respondent at different times as though they were from a monozygotic twin pair. As discussed above, this produces an overestimate of heritability. Had the estimated heritabilities been similar or lower than the values found in the original analysis, we would have been concerned that the findings might be due to chance. Instead, we found that the regression coefficients corresponding to heritabilities were very high, all exceeding 0.8. For nine of the 11 measures, the estimated heritability was close to unity, ranging from 0.934 to 1.036. For Instrumental Values of Children, for which the original estimate of heritability is only 0.034, the estimate in this sensitivity analysis was 0.825. Surprisingly, for Parental Stress, for which the original estimate of heritability was the highest of any measure, 0.724, this sensitivity analysis produced a relatively low heritability estimate of 0.884. While this analysis alone does not provide much evidence for heritability, the heritability values being known to be overestimates, it does support the values from the original analysis as plausible.

3. SUMMARY

Using the ACE model, we found strong evidence of heritability for the main scale of Negative Childbearing Motivation (NCM), for one of its subscales, and for one of the subscales of Positive Childbearing Motivation (PCM). We also found moderate evidence of heritability for three other subscales and weaker evidence for the main PCM scale and two of its subscales. Using the ADE model for two scales with evidence of nonadditivity and the AE model for the remaining scales, we found even stronger evidence of heritability. Only one of the eleven scales and subscales showed negligible estimated heritability. Despite the limitations of the sample, the results were robust to sensitivity analysis.

REFERENCES

Cherny, S.S., Cardon, L., R., Fulker, D.W., & DeFries, J.C. (1992). Differential heritability across levels of cognitive ability. *Behavior Genetics, 22,* 153-162.

Cherny, S.S., DeFries, J.C., & Fulker, D.W. (1992). Multiple regression analysis of twin data: A model-fitting approach. *Behavior Genetics, 22,* 489-497.

DeFries, J.C. & Fulker, D.W. (1985). Multiple regression analysis of twin data. *Behavior Genetics, 15,* 467-473.

Detterman, D.K., Thompson, L.A., & Pomin, R. (1990). Differences in heritability across groups differing in ability. *Behavior Genetics, 20,* 369-384.

Efron, B. (1981). *The Jackknife, Bootstrap and Other Resampling Plans.* Philadelphia, AP: Society for Industrial and Applied Mathematics.

Fuller, J.L. & Thompson, W.R. (1978). *Foundations of Behavior Genetics.* St. Louis: C.V. Mosby.

Haggard, E.A. (1958). *Intraclass Correlation and the Analysis of Variance.* New York: Holt, Rinehart & Winston.

Jackson, D.N. (1984). *Personality Research Form Manual.* Port Huron, MI: Research Psychologist Press.

Jöreskog, K.G., & Sörbom, D. (1996). *LISREL 8: User's Reference Guide.* Chicago, IL: Scientific Software International.

LaBuda, M.C., DeFries, J.C., & Fulker, D.W. (1986). Multiple regression analysis of twin data obtained from selected samples. *Genetic Epidemiology, 3,* 425-433.

Miller, W.B. (1986). Proception: An important fertility behavior. *Demography, 23,* 579-594.

Miller, W.B. (1992). Personality traits and developmental experiences as antecedents of childbearing motivation. *Demography, 29,* 265-285.

Miller, W.B. (1994). Childbearing motivations, desires, and intentions: A theoretical framework. *Genetic, Social, and General Psychology Monographs, 120,* 223-258.

Miller, W.B. (1995). Childbearing motivation and its measurement. *Journal of Biosocial Science, 27,* 473-487.

Miller, W.B. & Pasta, D.J. (1993). Motivational and nonmotivational determinants of child-number desires. *Population and Environment: A Journal of Interdisciplinary Studies, 15,* 113-138.

Miller, W.B. & Pasta, D.J. (1994). The psychology of child timing: A measurement instrument and a model. *Journal of Applied Social Psychology, 24,* 218-250.

Miller, W.B. & Pasta, D.J. (1995). Behavioral intentions: Which ones predict fertility behavior in married couples? *Journal of Applied Social Psychology, 25,* 530-555.

Miller, W.B., Pasta, D.J., MacMurray, J., Muhleman, D., Comings, D.E. (1999). Genetic Influences on Childbearing Motivation: Further Testing a Theoretical Framework. In J.L. Rodgers, D.C. Rowe, and W.B. Miller (Eds.), *Genetic Influences on Human Fertility and Sexuality.* Norwell, MA: Kluwer.

Rodgers, J.L. & Rowe, D.C. (1987). IQ similarity in twins, siblings, half-siblings, cousins, and random pairs. *Intelligence, 11,* 199-206.

Waller, N.G. (1994). A DeFries and Fulker regression model for genetic nonadditivity. *Behavior Genetics, 24,* 149-153.

Zieleniewski, A.J., Fulker, D.W., DeFries, J.C., & LaBuda, M.C. (1987). Multiple regression analysis of twin and sibling data. *Personality and Individual Differences, 8,* 787-791.

Notes

[1] The commentary at the end of this section by Udry raises questions about the gender composition and age differences in the pairs. Our sample had 46.2% different-sex pairs, 31.3% pairs of females, and 22.5% pairs of males. The median absolute difference in age was 2.7 years (mean 2.98), with 74% less than 4 years and 99% less than 7 years. When we performed analyses for same-sex pairs and for pairs with less than 4 years difference in age, we found the estimates of heritability were similar to or, in a few instances, somewhat larger than the full sample estimates.

GENETIC INFLUENCES ON HUMAN FERTILITY AND SEXUALITY: COMMENTARY ON CHAPTERS 1-5

J. Richard Udry

I use Hughes and Burleson's analysis of Fisher's Fundamental Theorem to provide guidance for the interpretation of themes in the preceding papers on the genetics of fertility and fertility motivation. Thank heavens Hughes and Burleson have helped us better understand Fisher's Fundamental Theorem. Without ways around the theorem, there would never be any genetic variance remaining to look at on traits important to fitness. The theorem says essentially that variance in genetic components of fitness will be wiped out quickly by selection. But I now propose that genetic components of fitness are not a stable set. As environments change, the components of fitness change, and new pools of genetic variance are exposed to selection.

Suppose that among early hunter-gatherers it was hard to stay alive long enough to reproduce, and only smart parents managed to rear offspring to maturity. Genetically controlled variance in intelligence would have plummeted, although one can look to Hughes and Burleson to imagine mechanisms maintaining variance. Today, it may be possible to be too smart, by deliberately depriving the future population of your genes. In these scenarios, intelligence is a fitness component in each period. Fisher's Theorem predicts two quite different effects on intelligence, trimming the bottom in one period and the top in the other period, but reducing its genetic variance in each case.

In a previous paper (Udry, 1996) I argued that when people have choices among behaviors, these behavior choices may express biological (here genetically influenced) behavioral predispositions. In different societies at different times, the choices and degrees of choice differ. It is easy to imagine that during human evolution the average person's dispositions toward childbearing had little influence on fertility. In the modern U.S. and other developed societies, childbearing motivations may have become a central determinant of completed fertility. If childbearing motivations, desires, and expectations have a genetic component, the individual variance in this genetic component was until modern times not trimmed much by selection, and was not a fitness component, because these motivations had no or slight effect on completed fertility or its timing. In modern societies it would have become a fitness component, for the first time opened up for genetic selection in entire populations.

Fitness components under strong selection pressures thus become strongly genetically determined, but they do not vary much in the population. Without population variance, the measurement of this genetic determination is not accessible by the methods used in the papers in this section, since nearly everyone has the same controlling genes after generations of strong selection. So not only are identical twins identical, but so are fraternal twins, (as well as with everyone else) with respect to the controlling genes. Another way of saying this is that at the controlling genetic loci, there are not different alleles with different effects. So neither DNA analysis nor behavior genetic analysis is a useful tool in this situation.

These issues become salient in each of the papers in this section. Kohler and Christensen study twins from Danish birth registers between 1910 and 1923, testing by behavior genetic design the genetic influence on fertility. They find that for cohorts born after 1920 there is a significantly greater similarity in the fertility of monozygotic than dizygotic female twins, and a significant birth cohort increase in the strength of the genetic effect.

Kohler and Christensen interpret this as an increase in the genetic effect on fertility for females. They believe this is because of the increase in domestic bargaining power of females after World War II. Expanding their argument, the increase in domestic bargaining power of females after WWII created the opportunity for females to exercise choice in fertility on the basis of their fertility-related motivations -- motivations that have a genetic foundation. This presumes that females had these motivations and the technical mechanisms for realizing them in fertility before WWII, but needed a shift in bargaining power toward women to make it happen. But if later cohorts were more effective contraceptors, this might also have caused the heritability of fertility to increase because women were more able to achieve their fertility desires.

If there was a shift in bargaining power away from men, why did this not cause a decline in the heritability of fertility among males in the late-born cohorts? Because the heritability estimates for males and females do not necessarily change in a zero-sum way. Each of these estimates can be changed independently by changing environmental factors.

Following the language of Hughes and Burleson, the individual differences in female reproductive motivation have become a fitness component. Genetic variance in female fertility motivation now affects actual fertility, and is therefore newly exposed to selection. Had it always been so exposed, it would have no genetic variance by now. This mechanism of maintaining genetic variance seems to fall under Hughes' and Burleson's mechanism 3.4—spatial and temporal variation in the environment, as a temporal variation. Yet it doesn't sound very much like their examples.

Pasta and Miller use a behavior genetic design to examine the heritability of childbearing motivations, using a DF approach. Their sample is quite unusual. The paired respondents are of unreported sex composition, so we do not know the proportion of same-sex pairs. Since I assume that the genetic determinants of fertility desires are different by sex, this should reduce the estimates of heritability. The pairs are also of unreported age composition. I would presume that age differences would complicate finding significant results. Many are of unknown degree of shared environment. The "cloning" strategy that creates "identical twins" out of single individuals will seem to many readers to involve unresolved assumptions.

In spite of (or perhaps because of) these methodological weaknesses, statistically significant and substantial heritability was determined for both positive and negative childbearing motivations. This says nothing about whether childbearing motivations are a fitness component, since we don't know from this research whether these motivations are correlated with actual fertility. However, there is an extensive literature that shows motivations affect fertility. Taking the findings at face value, and ignoring the paper's methodological quirks, heritability of motivations as high as their estimates plus the correlations of motivations with fertility would certainly start wiping out genetic variance in motivations. However, we might not notice any loss of variance even if we were looking at candidate genes, since as Hughes and Burleson argue, most variance is maintained by mutations.

The paper by Miller et al. constructs a brain-based theoretical system in which to embed the genetics of fertility motivation. By means of a structural equations model it tries to trace the pathways through which three neurotransmitters affect childbearing intentions. The intermediate way-stations are personality traits and childbearing motivation. This is a creative and useful way to think about the genetics of fertility. Some substantial correlations are found between the direct measures of specified polymorphisms and childbearing motivations. This will certainly encourage others to go for the DNA of motivations. If the structural equations models are specifying the steps in the theoretical process, they illustrate how far we have to go in connecting molecular genetics to childbearing. In Figure 3 of Miller et al., there are five steps from neurotransmitter receptors to fertility (the step from intentions to fertility is not available for Figure 3). There are two steps from receptors to PCM and NCM. Multiplying the coefficients across the paths gives you the predictive values. While a surprising number of coefficients in Figure 3 are statistically significant, the multiplied coefficients indicate that only a slight predictive value remains across two steps. This indicates that the receptor polymorphisms control only a small amount of variance in childbearing motivations when viewed through this diagram of the theory—not nearly as much as you might expect from the coefficients in the Pasta and Miller analysis. However, we should not be discouraged. Most researchers take it on faith that many different genes control any complex behavior.

Is the genetics of fertility and fertility motivation the same for males and females? The answer must be no for strictly physiological components of fertility. Tables 7, 8, and 9 in Miller et al. give ambiguous answers for fertility motivation. While the male and female coefficients for NCM and PCM are not significantly different, they are not very similar using the eyeball test. Kohler and Christensen find enough sex differences in their behavior-genetic models of fertility to reinforce our intuition that the strength and sources of genetic influence on fertility should differ by sex. Behavior-genetic research could offer guidance to molecular genetic approaches, and encouragement to Miller et al., as well as to Rodgers and Doughty, to articulate different models of fertility motivations by sex.

Rodgers and Doughty devote more attention to Fisher's Theorem than any of the other empirical papers. The implication of Fisher's theorem is that, with the variance maintenance mechanisms discussed by Hughes and Burleson excepted, the genetic variance in a fitness component should be selected away. Rodgers and Doughty suppose that "If fertility expectations are not linked at all to fertility outcomes, then by definition there should be low or no genetic variance in fertility expectations." But this stands Fisher's Theorem on its head. Their statement implies that if fertility expectations are NOT a fitness component, they will have no genetic variance. Surely they do not mean this. Not all genetic variance is related to fitness. Polymorphisms arise from many evolutionary processes, and are retained except as they become fitness components. For example, a genetic foundation for fertility motivation could arise as a genetic spandrel, be passed genetically from parents to children, and still have no effect on the number of offspring.

Eventually Rodgers and Doughty arrive at a reasonable assumption, in spite of their questionable logic: There can be any relationship between fertility expectations (or other fertility motivations) and actual fertility. They expect the relationship will be weak. Their expectation is that fertility motivations will be moderately heritable, and fertility itself less heritable. Their ingenious application of a behavior-genetic design to the NLSY data leads them to find that about half the overall variation in fertility expectations is genetic, and that at least part of the genetic variance in expectations becomes variance in fertility outcomes. Thus they conclude that small genetic influences appeared likely on fertility outcomes.

But this conclusion is an interpretation based on application of behavior genetic methods. If they had found NO genetic influence on completed fertility using their method, would this mean that fertility is not genetically influenced? Not at all. Behavior genetic analysis tells us whether the VARIANCE in fertility within a population is controlled by VARIANCE in genes in that population. If the answer is NO, then the genes controlling fertility are the same for all individuals. We can still investigate what genes those are through experimental molecular biology by deleting or substituting genes, or by watching for mutant genes. Weiner (1999) provides an engaging story of this research process (with simpler animals) that is intellectually accessible to the non-specialist.

I would like to see more comment in these papers on their implications for human evolution. Each paper raises the possibility that genetic fertility motivations are important in differential fertility. To the extent this is true, these motivations are evolutionary in their impact. The comments by Rodgers and Doughty on the genetic consequences of childlessness is important and cries out for further explication.

In summary we can reach the following conclusions from these papers and their implications.
1. Just because a behavior or preference is a component of the reproductive process does not mean it is a fitness component. Its genetic variance has to control fertility variance.
2. Fitness components also have genetic variance.
3. Fertility motivations are probably a fitness component.
4. Even if fertility motivations were not a fitness component now, they are of evolutionary interest because they might become a fitness component.
5. The papers in this section lead us to believe that further work on the genetics of human fertility and fertility motivation is likely to be fruitful. I especially look forward to more work on the molecular genetics of fertility motivations.

REFERENCES

Udry, J. R. (1996). Biosocial models of low fertility societies. In J. B.Casterline, R. D. Lee, and K. A. Foote (Eds.), *Fertility in the United States: New Patterns, New Theories. Supplement to Population and Development Review*, vol. 22.
Weiner, J. (1999). *Time, Love, Memory*. New York: Alfred A. Knopf.

Chapter 6

CONDITIONAL AND ALTERNATIVE REPRODUCTIVE STRATEGIES: INDIVIDUAL DIFFERENCES IN SUSCEPTIBILITY TO REARING EXPERIENCES

Jay Belsky

Key words: reproductive strategy, parental investment, pubertal timing, life-history, evolution

Abstract: Contrasting explanations of variation in reproductive strategies emphasize nature (i.e., heritability) and nurture (i.e., environment). In this chapter it is argued that both may be correct, but not simply because, as so commonly assumed, nature and nurture interact to shape development. Rather, the proposition is advanced that there may be variation in susceptibility to rearing influence. Thus, in the case of some individuals, early vs. late maturation, promiscuous vs. committed sexuality, producing many vs. few offspring, and low- vs. high-investment parenting may reflect heritable proclivities (i.e., alternative reproductive strategies). In other individuals, however, environmental effects may account for observed differences in such features of development and behavior that define reproductive strategies (i.e., conditional strategies). After advancing this differential-susceptibility argument with respect to reproductive strategy, evidence is reviewed suggesting that highly negative infants may be most susceptible to rearing influence, at least with respect to the development of problem behavior, self control, and conscience, in order to illustrate the argument that individuals may vary in the degree to which their development is shaped by forces of nature and nurture.

1. INTRODUCTION

In the middle decades of the 20th century, a good deal of research attention was paid to the phenomenon of father absence in response to Freudian theorizing emphasizing the Oedipal complex and thus the father's role in shaping the child's moral development. Indeed, one core extrapolation from Freud's theorizing about the Oedipal complex was that children growing up without fathers would have their moral development compromised. Such compromised development might take the form of breaking rules, engaging in promiscuous sexual activity, and being limited in the ability to establish and maintain intimate, heterosexual bonds. By the 1970s, interest in the phenomenon of father absence waned for a multiplicity of reasons. One, of course, was a general disenchantment with psychoanalytic theory. Also important, however, was inconsistency in the research literature on the effects of father absence and appreciation of the limits of many of the retrospective studies that had been carried out to test propositions derived from Freud's theory. Perhaps most important in this regard was the emergent recognition that the study of "social addresses" like father absence or social class were inherently limited because of their inability to illuminate processes of influence. Attention, developmentalists came to realize, needed to be paid to the actual experiences that children had in and outside of their homes. Simply characterizing a child as coming from a lower class household or one in which a father was present or absent afforded too little insight into *how* developmental processes operated. If we wanted to know what effects fathers—or mothers or other significant others in children's lives—might have upon development, the actual behavior of these significant others needed to be studied. Knowing simply whether a father was present or absent was insufficient because not all children inhabiting the same social address necessarily had similar experiences.

Despite psychologists' general abandonment of research that simply contrasted children growing up in father-absent homes with those growing up in families in which fathers were present, not all students of human development turned their backs on findings which derived from the research literature which was based on the social-address model. Most notable for purposes of this chapter was the work of two anthropologists, Patricia Draper and Henry Harpending, who, in 1982, published a quite original and provocative reinterpretation of the research literature on father absence. What was especially novel about the Draper and Harpending (1982) paper was their use of the construct of reproductive strategies, drawn from the field of behavioral ecology and life history theory, which emphasized the relative emphasis that organisms placed on mating and parenting, to organize data pertaining to interpersonal behavior in heterosexual relationships, sexual activity, and parenting. Life history theory assumes that selection will tend to favor phenotypic mechanisms that allocate limited resources to survival, growth and development, and reproduction in the way that is optimal for maximizing number of descendents. This necessarily involves tradeoffs among the components of fitness, because resources allocated to one component cannot be allocated to another. Resources allocated to survival, for example, cannot also be allocated to growth and development, nor can those allocated to the production of offspring also be allocated to their rearing.

Central to Draper and Harpending's (1982) analysis of the effects of father absence on psychological and behavioral development in adulthood was the proposition that such early family experience shaped the ways that individuals went about doing that which evolution designed all life forms to do, namely, reproduce. Fundamentally, these scholars contended that father absence fostered a reproductive strategy that emphasized mating over parenting. Thus, children growing up in father absent homes were inclined to mate early and promiscuously and invest little in parenting, whereas those growing up in father-present families delayed mating, established enduring pair bonds when they did mate, and then invested heavily in the rearing of a limited number of children.

What these theorists failed to specify, however, was the issue that had become central to the thinking of developmental psychologists—namely, the processes by which such proximate developmental mechanisms shaped reproductive behavior. That is, the question of *how* does father absence come to shape reproductive strategy was not of much concern to these anthropologists. *Why* father absence would shape mating and parenting was clear from the evolutionary arguments which Draper and Harpending (1982) advanced, but the developmental mechanisms by which early experience would shape reproductive behavior was never clear—nor apparently of much concern to these scholars.

My own training in developmental psychology made this a central issue for me, even though such a background had never alerted me to the possibility that linkages between childhood experience and adult functioning might have something to do with evolution. Beyond some limited exposure to the work of developmental behavior geneticists, my training in the 1970s, just like that of many developmentalists to this very day, hardly dealt with evolution at all. In fact, in only one respect beyond behavior genetics had evolution even been a category in my intellectual development. And this involved John Bowlby's (1969) theory of attachment, his notion of the environment of evolutionary adaptation that many developmentalists had come to deride, and the theoretical proposition that attachment behavior in the infant was selected because it promoted survival.

Given this personal intellectual history, it is thus not surprising that it was to Bowlby's (1969) theory of attachment that I first turned as I endeavored to think through the proximate "how" question that Draper and Harpending's (1982) provocative paper on father absence and reproductive strategy stimulated me to think about. Indeed, in two brief papers co-authored with Draper, I theorized that father absence, as well as a variety of early experiences in the family, might shape future reproductive strategy by influencing the young child's feelings of security and subsequent interpersonal relationships (Belsky, & Draper, 1987; Draper & Belsky, 1990). Such thinking reflected an attempt to put proximate psychological meat on the evolutionary bones that Draper and Harpending (1982) had so clearly articulated. But what became apparent as I thought through the issues at hand was that, at least from a developmental and psychological perspective, our initial theorizing lacked parsimony. That is, there was no shortage of psychological theorizing, some of it dating back at least to Freud, that already generated many of the same predictions that Draper and Harpending's thinking did, but without calling upon notions of reproductive strategy or of evolution more generally. A variety of social learning models, as well as attachment theory, advanced predictions that experience in the family during childhood, including growing up in a father-absent home, would affect psychological and behavioral development, including pair bonding and parental investment—though not in such explicitly evolutionary terms. Indeed, an abundance of correlational evidence already suggested that how individuals behaved in marital and even dating relationships in adulthood was related to how their own parents got along as husband and wife during their childhoods, and that how individuals parented as adults was related to how they had been reared as children. Even though there remained many questions about whether such correlational findings were artifacts of the retrospective methods used in many studies to measure marital and parent-child relations in the family of origin, and about whether the evidence reflected behavior-genetic effects masquerading as environmental ones, it seemed that positing an evolutionary rationale for such intergenerational linkages was simply not particularly parsimonious. Simply put, why invoke a more complex explanation dealing with natural selection and, specifically, reproductive strategy, when a less complex one dealing only with proximate attachment and social learning mechanisms might due the trick?

2. TOWARD AN EVOLUTIONARY THEORY OF SOCIALIZATION

When I raised these issues with my co-author, it became apparent that we were speaking quite different disciplinary languages. As far as the anthropologist Draper was concerned, there was simply no need to justify the invocation of evolutionary theory; it was a given. Ignoring it was what had to be justified. But for someone like myself, trained in the standard social science model, there was a huge intellectual problem to be confronted. However attracted I was to the notion of reproductive strategy and ultimate function, when it came to wrestling with Draper and Harpending's (1982) ideas linking father absence in childhood with mating and parenting in adulthood, or with relations between experience in the family of origin more generally and interpersonal functioning in adulthood, it was not clear that a sophisticated evolutionary theory generated predictions any different from a more simple, psychological one. So why, then, invoke the more sophisticated and seemingly less parsimonious explanation? What one needed, it seemed to me, was a unique and, better yet, unusual prediction derived from evolutionary thinking that, if confirmed, could not simply be assimilated by a more proximate, psychological theory. It was thus a critical test of an evolutionary perspective that I found myself pondering. Fortunately, one came to me while teaching a graduate seminar on family relations one evening. And that was that in addition to shaping mating and parenting behavior, particular kinds of experience in the family might also affect somatic development, specifically, the timing of puberty.

More specifically, I reasoned that a reproductive strategy emphasizing mating over parenting which would produce, at least in the environment of evolutionary adaptation, more progeny who were poorly cared rather than few who were well cared for, would be characterized by earlier maturation, as well as by more opportunistic mating with more sexual partners and less investment in parenting. Furthermore, I theorized that such a reproductive strategy would be stimulated by less sensitive and nurturant experiences in the family, which would themselves be fostered by stressful environmental conditions and promote attachment insecurity and behavioral difficulties. I reasoned, relatedly, that a reproductive strategy that resulted in the bearing of fewer children who were well cared for would involve later maturation, more mutually-beneficial mating with fewer sexual partners, and greater investment in parenting. Such a reproductive orientation, I further reasoned, would be stimulated by more sensitive and nurturant experiences in the family, which would themselves be fostered by benign if not supportive environmental conditions and promote attachment security and psychosocial competence.

What needs to be made crystal clear about this thinking was that while virtually every postulate advanced could be accounted for by prevailing psychological theories of development, be they social-learning or attachment derived, this was most certainly not the case with respect to pubertal-timing postulates. That is, these standard social science theories would simply never generate—nor could they even easily accommodate—a prediction pertaining to the timing of puberty. Here then was a unique and even unusual prediction that seemed consistent with an evolutionary-oriented reproductive-strategy theory of development which could not be derived from, or accounted for by, traditional developmental and standard social science model theories. To me this was critically important, as the philosophy of science that I had been schooled in stipulated that a good theory raises new questions, generates new research and findings, while accounting for old evidence and explanations. The evolutionary theory of socialization that Belsky, Steinberg and Draper (1991) eventually offered seemed to hold the promise of doing just that.

But theory, of course, is one thing and data quite another. Thus, the question now became one of evidence. Was there any empirical data consistent with the unique prediction I had generated linking variation in rearing experience with pubertal timing and which Jerome Barkow (1984) had also advanced, unbeknownst to me, in a footnote to a 1984 paper. In the field of developmental psychology there was essentially no pertinent data on the issue at hand because pubertal timing was principally regarded as an independent variable in that field. Indeed, ever since the 1920s evidence had been amassed documenting relations between earlier age of maturation and a host of psychological and behavioral outcomes including ones focussed upon problem behavior and sexual activity. Yet never had developmentalists conceptualized such sequelae of early maturation in terms of reproductive strategy, nor had it occurred to them that pubertal timing might be conceptualized as much as a dependent construct affected by earlier experiences as an independent construct that explained things that occurred later in development.

The first hint that the unique pubertal timing prediction might have some validity came from the work of Michelle Surbey (1990) who found that life-event stress predicted earlier age of menarche, as did father absence, in a sample of Canadian girls. Then Steinberg (1988) examined, in direct response to my theorizing, his own short-term longitudinal data on a large sample of young adolescents and found that daughters—but not sons—whose mother-child relationships could be characterized as more distant and less close matured somewhat earlier than other girls (Steinberg, 1988). As he was quite aware as someone specializing in family relations and adolescent development, no developmental psychologist had ever before been in a position to examine effects of family processes on pubertal timing, as prevailing thinking had only led to asking questions about the reverse process—that is, of the effects of pubertal timing on parent-child relations.

But to me the most promising findings derived from a direct test of the puberty prediction using a unique, prospective, longitudinal data set gathered in New Zealand of an entire birth cohort of more than 1,000 (half girls) born in the South Island town of Dunedin in 1972 and 1973 who had been studied every other year of their life since the age of three. It showed that both exposure to greater amounts of family conflict and rearing in a father-absent household independently predicted earlier age of menarche (Moffitt, Caspi, Silva, & Belsky, 1992). Even though the power of prediction was at best modest, the fact that the results were consistent with my theorizing were sufficiently encouraging that I concluded that I could proceed to write up my theoretical ideas and submit them for publication, which I did. They appeared in the developmental psychology journal *Child Development* in 1991(Belsky et. al., 1991).

Even though the evidence I had either come across—or stimulated the examination of—was intriguing, there could be no doubt that it was still limited from a theory-testing perspective. Because the data from all the aforementioned studies were correlational in nature, even when longitudinal, explanations other than our theory existed to account for them. Indeed, Moffitt was inclined to interpret results linking family conflict and father absence with pubertal timing as a function of genetics rather than experience. We were certainly aware of evidence that pubertal timing was heritable, so this alternative explanation could not be ignored. In fact, in the original theory paper published in 1991, we were explicit in this regard, titling one subsection A Strong Inference Tests of the Theory and arguing that behavior-genetic or experimental-intervention work was required before it could be concluded that early experiences in the family were, in fact, affecting pubertal timing as some correlational evidence had begun to suggest.

Since the publication of our theory, additional evidence has been reported. Perhaps most noteworthy is the work of Graber, Brooks-Gunn, and Warren (1995). What makes this research especially important is that Brooks-Gunn was a skeptic originally, having generated much data showing that extreme nutritional and exercise stress actually operated in a manner exactly opposite to that which we theorized; namely, it delayed rather than accelerated puberty. But as we noted originally in our theory, as had Surbey (1990) had in her own writings, there may be a need to distinguish between the kind of extreme stress that leads an organism to shut off maturation in order to survive and less extreme stress which may stimulate an organism to accelerate maturation given the prospect of an uncertain and risky future in which delayed maturation could translate into no reproduction at all.

To their surprise, Graber, Brooks-Gunn and Warren (1995) found that depression, as well as family conflict, predicted earlier rate of maturation in girls, just as Belsky, Steinberg and Draper (1991) had theorized. This depression finding was important in terms of our theory because we had posited that depression might be a psychological vehicle for mediating the effect of contextual and family stress on somatic development and because we looked for such a link in the New Zealand study but had not discerned it. As Brooks-Gunn noted upon first reporting their results at a conference well before the publication of their paper, the evidence linking depression in middle childhood with earlier age of menarche raised additional questions about the endocrinological processes that might be responsible for the effects discerned.

But the news on all fronts has not been positive for the Belsky, Steinberg and Draper (1991) theory. In addition to not all the links in the theoretical chain being supported in the New Zealand inquiry, it is also the case that other work has failed to substantiate the critical hypothesis linking rearing experience with pubertal timing. Of course, it is always difficult to know exactly how to deal with mixed evidence, as it can emerge for a variety of reasons. What may be called for eventually is a meta-analysis which collates and appraises all relevant evidence while examining potentially moderating factors such as methods used, sample sizes, and sample characteristics.

Beyond recent empirical evidence, it is important to note that some have criticized the theory on conceptual grounds, or by pointing to older evidence that seems inconsistent with it. Some have argued, for example, that it makes more sense in the face of environmental stress and resource uncertainty for organisms to delay maturation rather than accelerate it, and thereby wait for a more propitious time to "go to seed", so to speak (Miller, 1994). Others have suggested that there would be little bio-logic in an organism committing itself to a reproductive strategy based upon childhood experiences rather than remaining open and flexible so as to behave as ideally as circumstances permit at the actual time of mating and of parenting. Finally, Kevin MacDonald (1997) has drawn attention to historical data that indicate that in times of stress reproduction decreases rather than increases, at least at the level of populations, which also seems inconsistent with what our theory would predict.

Personally, I find all of these arguments—and the counter arguments that come to mind—intriguing. Even though I am not yet convinced that the theory is wrong, I am by no means convinced that it is right. What I am convinced of is that it has proven useful in at least stimulating some, though not many, developmentalists to think about pubertal timing as something that might be affected by earlier experiences, not just something that affects later development. I must say, however, that three trips to biennial meetings of the Society for Research in Child Development since publishing our ideas, as well as continued reading of mainstream developmental journals, disappoints and frustrates me because so few of my developmental colleagues have come to think in terms of reproductive strategies. Most continue to study problem behavior, pubertal timing, sexual activity, and parenting, even generating findings more rather than less consistent across these diverse developmental domains regarding the effects of a variety of factors, yet generally seem unwilling to even consider the idea that these might be component parts of a larger whole—a reproductive strategy— whether affected by experience or genetics or both.

In certain respects one can read our theory as old wine in a new bottle. That is, even though it is written in the evolutionary language of reproductive strategy, as well as of facultative responses to environmental conditions, and generates a unique and even unusual prediction regarding pubertal timing, it remains, in many respects, true to my standard social-science training. That is, even though I regard it as a nature-based theory of nurture, it remains an argument about environmental influences on development. Indeed, that is why it was subtitled "An evolutionary theory of socialization". After all, it attempts to account for why variation in rearing experience shapes some reproduction-related aspects of psychological and behavioral development, as well as how such developmental processes may operate.

3. ALTERNATIVE VS. CONDITIONAL STRATEGIES

In contrast to our theory of environmental influences, though certainly not in response to it, Rushton (1996) has offered an alternative theory of the origins of variation in reproductive strategy that seeks to explain a host of racial differences. Without going into detail, suffice it to say that he would probably argue, as would most behavior geneticists, that many of the effects that our theory attributes to variation in rearing experience, should be attributed to genetic differences. That is, rearing differences, differences in psychological development, differences in pubertal timing, and differences in mating and parenting, that is, in reproductive strategy, come to be correlated not because of any true environmental influence, but rather because common genes affect all these components of our theoretical model. Personally, I have no doubt whatsoever that there is some, even a large amount of truth, to this argument. But, to adopt the terminology employed by Rowe, Vazsonyi and Figueredo (1997) and others to demarcate the distinction just made, I do not regard a *conditional* reproductive strategy perspective, that is, one which emphasizes environmentally-triggered processes, and an *alternative* reproductive strategy framework, that is, one which emphasizes genotypic differences, as mutually exclusive. Indeed, in our original theoretical formulation, even though we clearly emphasized the environmentally-triggered, conditional-strategy approach while acknowledging the genotypically-determined, alternative strategy approach, we explicitly raised the prospect of a hybrid formulation. And, in the years since we published our original theoretical piece, I have become ever more intrigued by the prospect that in some cases reproductive strategies are made, just as we theorized, whereas in other cases they are born, as Rushton, Rowe and other behavior geneticists have argued.

In the same way, then, that Linda Mealey (1995) has argued in favor of a distinction between primary and secondary sociopaths, that is, between those destined by their genotypes to become sociopaths and those made into sociopaths by their developmental experiences, I think it is useful to think of the two types of developmental trajectories outlined in our theory in similar terms. Indeed, it seems to me that there would be reproductive payoffs in parents bearing offspring who varied in terms of their susceptibility to rearing influence and thus, some whose traits and developmental trajectory are more or less fixed and others whose traits and trajectory are more or less plastic. Would not such a diversified approach to bearing offspring serve to hedge bets given an ever-present uncertain future? The payoff for a fixed type in terms of reproductive success would emerge under conditions in which there turned out to be a very good match between the nature of the fixed trajectory and the developmental context. Thus, a secure, empathic, trusting individual genotypically predisposed to delay sexual maturation and activity, and to establish enduring and trusting pair bonds would likely thrive in a context which values and rewards such an orientation. In contrast, in the opposite kind of context, the insecure, mistrusting, advantage-taking individual genotypically predisposed to mature early and establish multiple and opportunistic sexual liaisons while investing little in parenting would likely thrive, at least relative to his more trusting counterpart. Needless to say, when such fixed types encounter ecological niches into which they prove to be poor fits, average payoffs for these types should be greatly reduced.

More plastic individuals, whose psychological and behavioral development, and thus reproductive strategy, is highly susceptible to rearing influence should, instead, be more capable of fitting a wider range of environments. Rather than being a specialist prepared by genetic endowment to thrive in one particular niche, the more plastic individual should be reactive to a range of rearing milieus and thus, in theory at least, capable of fitting a variety of niches, depending upon which his developmental experiences have prepared him for. Even though such persons may never fit any niche as well as a fixed type who finds him or herself in the context for which she or he was selected to thrive, the increased range of niches that the more plastic type could flourish in would compensate for the less than perfect fit that might be achieved in any one of them.

Needless to say, developmental varieties might be more complex than this simple formulation presupposes. Not only might it make sense to think in terms of more than two prototypic reproductive strategies and related developmental trajectories, but it might prove useful to think in terms of some individuals being more vs. less susceptible to developing in certain ways rather than in all ways. One can imagine, for example, an individual who is genotypically-inclined to be only moderately opportunistic, but who could, under particularly harsh rearing conditions, become extremely so, even though he would be far less susceptible to the influence of more benign or benevolent rearing conditions. Relatedly, another individual might be genotypically-inclined to be only moderately secure, trusting and helpful, but under especially nurturant or supportive rearing conditions, could become extremely so, even though she would be far less susceptible to the effects of harsh rearing conditions. In other words, among the more plastic types of individuals that I am imagining, some might be genotypically predisposed to develop in some ways rather than in others given particular rearing conditions.

4. DIFFERENTIAL SUSCEPTIBILITY TO REARING INFLUENCE

Clearly, in this more complex formulation of the role that rearing conditions might play in shaping reproductive strategy, a standard-social-science-model belief in the influence of rearing experiences on psychological and behavioral development and, thereby, reproductive strategy is retained. What is being suggested, however, is that individuals may vary for heritable (i.e., genetic) reasons in their susceptibility to such rearing influence. And, moreover, that it may be reproductively strategic for parents to bear progeny who vary not only in their psychological and physical traits, but in their very susceptibility to being affected by the nature of the rearing to which they are exposed.

However attractive may be the conceptualization just advanced of differential susceptibility to rearing influence—especially to those who retain some allegiance to the standard-social-science model—it is not unreasonable to wonder whether we can move beyond theoretical formulation and ask, fundamentally, whether any data do indeed indicate that children vary in their susceptibility to rearing. As will become apparent soon, an emerging body of evidence does, in fact, point in this direction and indicates, moreover—and perhaps somewhat surprisingly—that it is infants with negative and perhaps even difficult temperaments who may be especially susceptible to rearing influence. The examination of some pertinent developmental data begins with a brief summary of some pertinent primate data before proceeding to consider that obtained from studies of human infants, toddlers and preschoolers.

The first work to share involves the study of rhesus macaques who have been selectively bred by Steve Suomi (1997) and his coworkers at the National Institute of Child Health and Human Development so as to vary in their fearfulness and proclivity to become anxious. "Up-tight" is the colloquial term that Suomi uses to refer to those rhesus monkeys selectively bred to be highly anxious and fearful. Among the wide variety of experiments that Suomi and coworkers have carried out using these up-tight monkeys, the one of particular relevance to this chapter involves the cross-fostering of these highly-anxious monkeys and their counterparts (selectively bred to fall in the more normal range of fear and anxiety) to two different groups of foster mothers. One group consists of especially skilled and highly nurturant mothers and another group manifests just average levels of skill and nurturance. To be noted is that these foster mothers were classified as highly skilled or just averaged based upon the extent to which they had groomed their own offspring and the degree to which they had rejected them during the course of weaning. Those foster mothers classified as highly skilled evinced especially high levels of grooming and especially low levels of rejection when rearing previous offspring, a pattern of behavior which they maintained when caring for their foster children.

What is especially interesting about one series of studies that Suomi and colleagues carried out is that the effect of rearing condition was rather dramatic in the case of one group of the selectively-bred monkeys, but not in the other (i.e., treatment X aptitude interaction). Even though being reared by an extremely skilled or a merely average mother exerted only limited impact on the average infants, it exerted a marked effect on the up-tight ones. Whereas up-tight infants foster-reared to average mothers exhibited expected deficits in early exploration patterns and exaggerated bio-behavioral responses to minor environmental perturbations, these same high-reactive infants actually appeared to be behaviorally precocious when cross-fostered to nurturant females. Indeed, these latter infants physically separated from their mother at an earlier point in development, locomoted and explored their environment more, and displayed less behavioral disturbance during weaning than not only the high-reactive infants cross-fostered to average mothers, but even the average infants reared by either type of foster mother. Follow up investigation of these selectively bred and differentially foster-reared monkeys when they were moved into larger social groups at six months of age revealed additional evidence of differential susceptibility to rearing influence, marked by seemingly positive outcomes for the up-tight monkeys reared by particularly competent foster mothers. These individuals became especially adept at recruiting and retaining other group members as allies in response to agonistic encounters and, perhaps as a consequence, they subsequently rose to and maintained top positions in the group's dominance hierarchy. In contrast, temperamentally-similar, high-reactive individuals cross-fostered to control mothers, tended to drop to and remain at the bottom of the same dominance hierarchy. Importantly, no such rearing effects were evident among the average infants. In other words, the range of reaction of the up-tight monkeys in response to these contrasting rearing conditions greatly exceeded that of their average-reactive counterparts subjected to the same variation in rearing regimens.

Having considered the question of differential susceptibility to rearing influence in the case of rhesus macaques, it is now time to turn attention to the human condition. The first work to be discussed comes from a series of studies carried out by Grazyna Kochanska (1993) of the University of Iowa testing the hypothesis that more fearful, inhibited and negatively-emotional (i.e., Up-tight) children would be more affected by maternal discipline and socialization than would less negative ones when it came to the development of self control. The subjects of this work were 103 toddlers age 26-41 months whose affective responses to a series of mildly threatening events were studied in order to evaluate their fearfulness. The so-called risky events to which these toddlers were exposed included an invitation to drive a little vehicle decorated with animal masks, plants, disembodied dolls heads and odd birds; opportunities to jump on a trampoline, reach inside a big black box to retrieve a hidden toy, and try on a gorilla mask; as well as the chance to interact with an unfamiliar female dressed in a clown suit who was wearing a bright red wig. Needless to say, some children were more distressed by these novel and disconcerting events than were others. In addition to observing children's reaction to these mildly arousing events, mothers completed questionnaires assessing the extent to which their child was fearful and anxiety prone. Thus, both observations and maternal reports were employed to assess early temperament.

In order to assess the rearing experiences of these children, multiple methods of measurement also were utilized. Thus, mothers were observed interacting with their toddlers during a toy clean up session and rated in terms of the extent to which their management strategies involved gentle guidance rather than forceful, power-assertive control. And maternal reports concerning the disciplinary strategies used to manage toddler misbehavior were also obtained. Finally, to measure the extent to which children internalized family standards of proper behavior, children were observed while alone in the lab room. Kochanska wanted to see whether, after being left alone to work on a rather boring task, the children would misbehave and play with a set of attractive toys which they had been told not to touch but which were purposefully placed nearby to challenge them. In addition to these observational data, mothers were queried about their children's behavior, particularly the extent to which they followed rules and behaved in ways consistent with maternal expectations.

Results of analyses designed to determine whether maternal child rearing strategy—particularly the extent to which she relied upon gentle guidance vs. forceful control—predicted children's functioning differentially as a function of child temperament proved most interesting. Notably, the apparent effect of mothering varied dramatically across more and less negatively reactive children. With respect to refraining from playing with the off-limits toy in the lab when no adults were present, maternal behavior explained only 1% of the variance in the case the low fearful/anxious children, but a significant 23% in the case of the highly fearful children. With respect to the maternal report of children's proclivity to follow rules, maternal gentle discipline which de-emphasized power assertion accounted for 4% of the variance in the case of the low fearful/anxious children, but a significant 18% in the case of the highly fearful and anxious children.

Recently, Kochanska and her colleagues (1997) sought to determine whether these differential effects of mothering discerned during the toddler years were evident as early as the first year of life. Thus, in a more recent investigation she observed 112 8-10 month olds and their mothers in a situation in which it was the mother's job to keep the child from touching and playing with a colorful toy plant with lots of attractive things hanging from it. The behavior to be explained in this work was the child's probability of complying with a maternal directive not to touch the plant in the 5-second period following each maternal prohibition. Children's negative emotional responses to a variety of stressful situations were independently measured once again so that more and less negative infants could be distinguished. Once more mother's use of a gentle style of controlling the child that de-emphasized power was scored. As in the aforementioned work, though this time dealing with much younger children, Kochanska found that maternal discipline predicted the restraint of the more negative and fearful children but not of the less negative and less fearful children. In fact, in the case of the 25% of the sample that reacted most negatively to the fearful stimuli, maternal discipline explained a significant 41% of the variance in children's restraint; the corresponding figure in the case of the 25% least fearful children was an insignificant 2% of the variance.

One of the limitations of this work of Kochanska's is its cross-sectional design, with parenting, temperament and children's self-control being measured at roughly the same point in time. Three recent studies redress this limitation while specifically testing the a-priori proposition that infants with more negative temperaments may be more susceptible to rearing influence, at least insofar as the development of self control is concerned. One investigation carried out in Israel focussed upon difficult infant temperament, measured via maternal reports and observed negativity, synchrony in mother-infant interaction during face-to-face exchanges when infants were nine months of age, and self control at age two, as measured by compliance with maternal clean-up requests and capacity to refrain from eating candy until permitted to (Feldman, Greenbaum, Yirmiya, 1999). Just as expected, mutually synchronous mother-infant interaction predicted greater self control more strongly in the case of children who had had difficult temperament as infants ($r=.65$) than in the case of infants who did not ($r=.25$).

The second longitudinal test of the differential-susceptibility-to rearing hypothesis was carried out by my colleagues and myself as part of a larger investigation of some 125 working- and middle-class Caucasian families rearing firstborn sons in central Pennsylvania (Belsky, Hsieh, & Crnic, 1998). In this research, we measured infant negative emotionality at one year of age, relying upon a variety of procedures, including parental reports of temperament, reactions to unfamiliar adults and separations from parents, and responses to frustrating situations involving a parent taking away an attractive toy and refusing to return it or interact with the child. In order to measure childrearing practices, we visited the homes of these families when children were 15, 21, 27 and 33 months old and observed mother-father-child interaction. The paired one-hour visits at each age usually took place in the early evening, during which parents were instructed to go about their everyday household routine. Every 15 minutes a series of ratings of maternal and paternal behavior were made which focussed upon negative parental behavior and positive parental behavior. These 15-minute ratings were averaged across the four 15-minute observation epochs each day and then across the entire 8 days in order to create positive and negative mothering and fathering composite scores. Positive parenting reflected the extent to which mothers or fathers were positively affectionate with their child and sensitive in interacting with the child. Negative parenting reflected the extent to which the parent was hostile and irritable and intrusive in controlling the child.

With measures of positive and negative mothering and fathering in hand, we sought to predict two separate developmental outcomes when children were 36 months of age: how shy, inhibited and reticent the child was during a series of laboratory procedures and the extent to which mothers and fathers reported the child as having externalizing behavior problems involving aggression and noncompliance. Just as in the work of Suomi and of Kochanska, we, too, found that our capacity to predict children's functioning using measures of child rearing was noticeably greater in the case of children who had been highly negative as infants than in the case of infants who scored below the median on our measure of infant negativity.

A similar pattern of findings emerged when Deater-Decker and Dodge (1997) examined the behavior problems of children in middle childhood in response to my theorizing about differential susceptibility to rearing influence (Belsky, 1997). These investigators took advantage of retrospective maternal reports of infant temperament obtained just before children entered kindergarten and concurrent measures of parental use of harsh discipline to manage the child in order to predict children's externalizing problems measured when children were in kindergarten and grades 1, 2,3, 4 and 5. More specifically, they examined the mean correlation between maternal harsh discipline prior to kindergarten entry and externalizing problems during each of the first 6 years of school, separately for children who were and were not retrospectively characterized as having highly persistent temperaments as infants. Consistent with previous findings, they found that children, who were described as infants as highly resistant to intrusion and persistent in pursuing forbidden activities—that is, negatively reactive to efforts to control them—were apparently more affected by the discipline they received than were other children. Thus, the children who as infants were characterized as being hyper-resistant to control were more likely than others to show many or few externalizing problems depending upon the degree to which they experienced harsh punishment than were children who were less negatively responsive to efforts to control them. In other words, the effect of harsh punishment on externalizing problems presumed by many developmental adherents to the standard social science model seemed restricted to a subset of children. Once again, then, we see that all children did not appear equally affected by the quality of care they received, even when that care involved exposure to harsh discipline.

What is especially intriguing about the emerging evidence suggesting that highly negative infants are particularly susceptible to rearing influence is that it accords nicely with some work on the theorized role of maternal sensitive responsiveness in promoting infant attachment security. One well cited study among developmental researchers is that reported almost two decades ago by Crockenberg (1981), designed to test the hypothesis that the availability of social support to a mother would enhance her capacity to be sensitively responsive to her infant and, thereby, to facilitate the development of a secure rather than insecure infant-mother attachment relationship. In this work, Crockenberg (1981) did indeed find that social support was predictive of infant attachment security but only, quite intriguingly, in the case of highly irritable infants. Once again we see, then, that it is highly negative infants who may be particularly susceptible to rearing influence, in this case benefiting from the social support afforded their mothers.

The effect of maternal care on infant-mother attachment security has been extensively studied by developmental psychologists. To my knowledge, however, no one has specifically addressed the question of whether, as the work of Crockenberg (1981) might now be read to indicate, it is highly negative infants who are most affected by the quality of care they receive when it comes to explaining why some infants develop secure and others insecure attachments. Is it possible that this could explain why the theorized and repeatedly discerned relation between quality of mothering and attachment security turns out to be so modest in magnitude? In 1987, Goldsmith and Alansky (1987) reported a meta-analysis of the then available literature and chronicled a reliable effect size of .16 of the impact of mothering on infant attachment security. More recently, in a recently published up date of that meta-analysis, this time involving more than 4,000 infants and their mothers who participated in some 66 investigations, De Wolff and van Ijzendoorn (1997) chronicled an effect size of .22.

This modest relation between maternal sensitive responsiveness and infant attachment security derived from correlational studies is highlighted here because it contrasts so markedly with the findings of perhaps the most carefully conducted study to date examining the effect of sensitive mothering on attachment security. In this Dutch work, van den Boom (1994) randomly assigned 50 lower-class mothers to an experimental group that received an intervention geared toward promoting maternal sensitivity and 50 other mothers to a control group that did not receive any such intervention. Not only did the intervention enhance, as was intended, the sensitivity of mothers to their infants' positive and negative emotional cues, as measured by home observations of maternal behavior when infants were 9 months of age, but it also affected, as anticipated, infant attachment security, as measured three months later. In fact, whereas only 22% of the infants in the control group were classified as securely attached at 12 months of age, the corresponding figure in the case of the experimental mothers whose mothering had been enhanced, was three times greater--66%. Important to emphasize is not only that experimental and control mothers were randomly assigned to treatment condition, but that home observations before such randomization took place revealed that the two groups of mothers were no different in the quality of care they provided their 6-month old infants.

Why are these rather dramatic results of this carefully controlled experiment, which clearly chronicles a causal effect of mothering on attachment security, so at odds with the small effects discerned in the two meta-analyses of correlational studies? In light of the arguments advanced in this chapter, it seems especially notable that the answer to this question might well have something to do with the unique nature of the sample that van den Boom studied—a small detail intentionally omitted up to this point.

Purposefully, this Dutch investigator preselected her sample to consist of infants who were highly irritable as newborns! Indeed, in order to be included in either the experimental or control condition, infants had to score high in irritability on two separate assessments of neonatal behavior carried out within the opening days of life. Although there can be no certainty that it was the heightened irritability of the Dutch infants that made them so susceptible to improvements in maternal sensitive responsiveness, this interpretation certainly seems credible in light of all the results summarized through this point. Indeed, when these Dutch data are considered in the context of all the evidence presented, there would seem to be grounds to conclude that children do seemingly vary in their susceptibility to rearing influence and that one factor determining whether they are likely to be affected by care that is more vs. less nurturant is their temperament and especially their proneness to negative emotions.

5.   CONCLUSION

It is important in closing to return to the title of this chapter—conditional *and* alternative reproductive strategies—given the extended focus in the preceding section of the chapter upon the chapter subtitle: differential susceptibility to rearing influence. Ultimately, the primary goal of this contribution to this volume has not been to argue that highly negative infants are more susceptible to rearing influence than are less negative infants, though accumulating evidence certainly suggests that this may well be the case. The larger point to be made by drawing upon such developmental data is that the notion that children vary in susceptibilty to rearing influence is consistent with the idea which gave birth to it—namely, that we need to think in terms of conditional **and** alternative reproductive strategies. That is, we may err, as perhaps Belsky, Steinberg and Draper (1991) did, in over-emphasizing the role of the environment or, as others might, in over emphasizing heritability in shaping developmental trajectories and reproductive strategies. This is because it may prove to be the case that both conceptualizations are accurate, but for some individuals more than for others. In point of fact, it was not developmental data that stimulated my thinking about variation in susceptibility to rearing influence, but rather the other way around. Evolutionary theorizing led me to entertain the possibility that children, especially within a family, should vary in their susceptibility to rearing influence, as this would be yet another way that parents could hedge their reproductive bets in the face of an ever present uncertain future. Whereas it would make sense to produce some children who are fixed strategists, perhaps entirely immune to socialization efforts, who would thrive in particular contexts that fit their proclivities, it would also make sense to produce some who are more plastic, capable of fitting and thriving in a variety of niches depending upon the rearing conditions they encountered while growing up. Ultimately, however, it may not be best to think in terms of discrete fixed and plastic types, but rather in terms of a continuum of plasticity. Either way, though, the developmental landscape becomes more dynamic—and more empirically challenging—than it appears when we think exclusively in terms of conditional or alternative strategies. Whether or not reproductive strategies, which includes sexual behavior, will prove consistent with the arguments advanced—namely that for some individuals it will prove to be genetically determined whereas for others it will prove to be environmentally shaped—remains to determined. As should be apparent by this time, my prediction is that individuals vary—for heritable reasons—in their susceptibility to rearing influence. Thus, for some, variation in their sexual behavior will be principally a function of their genes, whereas for others it will be principally a function of their rearing experiences.

REFERENCES

Barkow, J. (1984). The distance between genes and culture. *Journal of Anthropological Research, 40*, 9-14.

Belsky, J. (1997). Variation in susceptibility to rearing influence: An evolutionary argument. *Psychological Inquiry, 8*, 182-186.

Belsky, J., & Draper, P. (1987). Reproductive strategies and radical solutions. *Transaction/Society, 24*, 20-24.

Belsky, J., Hsieh, K., & Crnic, K. (1998). Mothering, fathering, and infant negativity as antecedents of boys= externalizing problems and inhibition at age 3: Differential susceptibility to rearing influence? *Development and Psychopathology, 10*, 301-319.

Belsky, J., Steinberg, L., & Draper, P. (1991). Childhood experience, interpersonal development, and reproductive strategy: An evolutionary theory of socialization. *Child Development, 62*, 647-670.

Bowlby, J. (1969). *Attachment and loss: Vol. 1. Attachment.* N.Y.: Basic Books.

Crockenberg, S. (1981). Infant irritability, mother responsiveness, and social support influences on the security of infant-mohter attachment. *Child Development, 52*, 857-865.

Deater-Deckard, K. & Dodge, K. (1997). Spare the rod, spoil the authors: Emerging themes in research on parenting. *Psychological Inquiry, 8*, 230-235.

De Wolff, M. & van Ijzendoorn, M. (1997). Sensitivity and attachment: A meta-analysis on parental antecedents of infant attachment. *Child Development, 68*, 571-591.

Draper, P. & Belsky, J. (1990). The relvance of evolutionary thinking for issues in personality development. *Journal of Personality, 58*, 141-162.

Draper, P. & Harpending, H. (1982). Father absence and reprodcutive strategy: An evolutionary perspective. *Journal of Anthropological Research, 38*, 255-273.

Feldman, R., Greenbaum, C., & Yirmiya, N. (1999). Mother-infant affect synchrony as an antecedent of the emergence of self-control. *Developmental Psychology, 35,* 223-231.

Goldsmith, H. & Alansky, J. (1987). Maternal and infant predictors of attachment. *Journal of Consulting and Clinical Psychology, 55,* 805-816.

Graber, J., Brooks-Gunn, J., & Warren, M. (1995). The antecedents of menarcheal age. *Child Development, 66,* 346-359.

Kochanska, G. (1993).Toward a synthesis of parental socialization and child temperament in early development of conscience. *Child Development, 64,* 325-347.

Kochanska, G., Mordhorst, M., & Reschly, A. (5 April, 1997). *Child temperament and maternal discipline as contributors to emerging restraint in infancy.* Paper presented at the biennial meetings of the Society for Research in Child Development, Washington, D.C.

MacDonald, K. (1997). Life history theory and human reproductive behavior: Envionrmental/contextual influences and heritable variation. *Human Nature, 8,* 327-360.

Mealey, L. (1995). The sociobiology of sociopathy: An integrated evolutionary model. *Behavioral and Brain Sciences, 18,* 523-599.

Miller, E. (1994). Optimal adjustment of mating effort to environmental conditions. *Mankind Quarterly, 34,* 294-316.

Moffitt, T., Caspi, A., Silva, P., & Belsky, J. (1992). Childhood experience and the onset of menarche: A test of a sociobiological model. *Child Development, 63,* 47-58.

Rowe, D., Vazsonyi, A., & Figueredo, A. (7 June 1997). Mating effort in adolescence: Conditional or alternative reproductive strategies. Paper presented at the annual meetings of the Human Behavior and Evolution Society, Tuscon, AZ.

Rushton, J. P. (1996). Race, genetics, and human reproductive strategies. Genetic, Social, and General Psychology Monographs, *122,* 21-53.

Steinberg, L., (1988). Reciprocal relation between parent-child distance and pubertal maturation. *Developmental Psychology, 24,* 122-128.

Sulloway, F. (1997*). Born to rebel.* NY: Pantheon.

Surbey, M. (1990). Family composition, stress, and human menarache. In F. Bercovitch & T. Zeigler (Eds*.), The socioendocrinology of primate reproduction.* New York: Liss.

Suomi, S. (1997). Early determinants of behaviour: Evidence from primate studies., *British Medical Bulletin, 53,* 170-184.

Van den Boom, D. (1994). The influence of temperament and mothering on attachment and exploration: An experimental manipulation of sensitive responsiveness among lower-class mothers and irritable infants. *Child Development, 65,* 1457-1477.

Chapter 7

ENVIRONMENTAL AND GENETIC INFLUENCES ON PUBERTAL DEVELOPMENT: EVOLUTIONARY LIFE HISTORY TRAITS?

David C. Rowe

Key words: heritability, evolution, menarche, pubertal development

Abstract: Precocious pubertal development has been observed among girls in single parent families and among girls exposed to family conflict. One explanation for their precocious puberty is that it is evolutionarily adaptive (called here the "life history theory" view). Another explanation is that mothers simply pass on genes for precocious puberty. Thus, family environmental circumstances are not causally determinative of rates of pubertal development. The evidence for these two views was evaluated using the genetically-informative sibling pairs from the *National Longitudinal Study of Adolescent Health*. Both menarcheal age and pubertal timing were heritable (h^2 = .44 and .40, respectively), with negligible shared environmental variation. However, in White girls, greater parental warmth delayed puberty, as did living in two parent families. A test of genetic influences underlying the correlation of warmth and pubertal timing was negative, however. Although some of these findings were inconsistent with the evolutionary life history theory, they did not disprove it decisively.

1.   INTRODUCTION

Pubertal development is usually thought of as a developmental unfolding. As children enter into adolescence, their circulating hormones change. In response to these hormonal changes, sexually dimorphic physical traits appear. Boys will develop beards, a deeper voice, and greater upper body strength. They become capable of ejaculating sperm. Girls will develop breasts, wider hips, and begin their cycles of ovulation at menarche. No particular environmental event is needed to stimulate these maturational changes; they happen in all human children in about the same way in all cultures.

Although pubertal development is universal, extreme environmental conditions can alter the normal progression of puberty. Malnourishment can delay the onset of menarche in girls, or make pubertal cycling cease. In evolutionary terms, delay of puberty may enhance survival, so that reproduction occurs when nourishment improves. Extremes of physical exercise can also delay puberty (in dancers, see Calabrese et al., 1983). Pubertal delay, however, is reversible when girls decrease exercise or increase their caloric intake. Furthermore, starvation has wide systemic effects, with pubertal delay being only one manifestation. Thus, these extreme examples are not necessarily evidence that pubertal development is responsive to variation in more normal environmental conditions in western countries.

Evolutionarily-inspired studies have also revealed interesting and unexpected correlates of pubertal maturation with variation in the social and psychological influences of family environments. Age of menache is more precocious in girls living only with their mothers and in girls who experience more family conflict and less warmth from their parents (Ellis et al., 1999; Graber, Brooks-Gunn, & Warren, 1995; Moffitt et al., 1992; Steinberg, 1988). However, father absence may be less critical than the presence of unrelated men in the family; Ellis and Garber (in press) found that the duration of step-father presence was a stronger predictor of early maturation than the duration of absence of the biological father. The effect of stepfather presence was further intensified by conflict with the unrelated parent. The idea that a genetically programmed course of physical development can be altered by the emotions of family life, or by the mere presence of unrelated males, is a counter-intuitive prediction of the evolutionarily-based theories of life histories (Belsky, Steinberg, & Draper, 1991). The research design used in these studies of precocious puberty contain a potential methodological confound, however. This confound is that genes shared by parent and child may promote early pubertal maturation in both generations.

In the first section of this chapter, the evolutionary life history and behavioral genetic theories of girls' precocious puberty are contrasted (see also, MacDonald, 1997). In the second section, the competing predictions of these theories are evaluated using sibling data from the *National Longitudinal Study of Adolescent Health* (Udry & Bearman, 1998). These sibling data can answers some questions about precocious pubertal development in girls, but perhaps not untangle the thicket of causal influences, as discussed below.

1.1   Evolutionary Theory of Life Histories and Pubertal Development

A life history is the sequence and timing of important adaptive traits, such as the age of parent at first birth, birth spacing, and ultimate family size. Such traits are critical from an evolutionary perspective because they determine ultimate genetic representation in the next generation.   Life history traits always involve trade-offs because the optimization of one trait may be detrimental to another. For example, in birds a large clutch creates the potential of many surviving offspring, but if weather makes food scarce, the poorly fed offspring of a large clutch may fare worse than fewer offspring in a small clutch, with the latter parents having a better evolutionary outcome.   Evolution sculpts clutch size toward the optimal number of eggs laid for long-term survival under variable seasonal conditions.   Life history shows variability within populations. Conditional life histories are those that depend on environmental conditions to trigger one life history or another. As an illustration, consider that whether turtle eggs develop into male or female adults depends on the temperature of the sand surrounding them. Life histories can also vary within populations for genetic reasons.   Such genetically-influenced life histories are called "alternative" life histories. In this chapter, the term "life history" is used to refer to the conditional type of life history. Genetic effects on life histories, to the extent they are found, may serve as alternative life histories. In this chapter, however, we do not consider evidence of genetic influence on life history traits other than on pubertal development.

Although, as stated earlier, nutrition and exercise are recognized as moderators of pubertal development, more controversial evolutionary theories have emerged that assert that events early in family life also moderate pubertal development. One influential life history theory was advanced by Belsky and his colleagues in the journal *Child Development* (Belsky et al., 1991; see also Draper & Belsky, 1990). The "Belsky hypothesis" used speculations about evolutionary adaptations to advance hypotheses about typical pathways of child development. Not all the relations in these pathways are necessarily established. In the Type II pathway, the early environment is stable. Children are raised by their mothers and fathers. Their parents are loving and consistent in the application of discipline. In Judy Harris's (1998) terms, their discipline is "just right," neither too harsh nor neglectful. The boys from these families are low in aggression. The girls in these families are not anxious. They also are sexually restrained prior to marriage and form marriages that endure.

In the Type I pathway, the early circumstances of children is different. They typically grow up in single parent households, or in two parent households marked by serious physical or emotional conflict between their parents. Boys who experience this type of early environment (from birth to seven years) later become more aggressive and antisocial. Girls from the same types of families later become more anxious and sexually promiscuous; they do not form lasting relationships with men. From the vantage of widely accepted social values, higher levels of anxiety, sexual promiscuity, with its resultant pregnancies and unwed motherhoods in girls, and higher levels of crime in boys are undesirable outcomes. In addition to these outcomes, Belsky et al. added one further difference between the Type II and Type I pathways: girls in the Type I pathway would experience an earlier age of menarche, that is, a more rapid pubertal development, than girls in the Type II pathway. This hypothesis, that physical development would change in concert with psychological development, was a most novel prediction of their theory.

Belsky et al.'s (1991) explanation for variation in puberty and aggressive and sexual behavior relies on ideas from the evolutionary theory of life history traits. According to Belsky et al., in the Type II pathway offspring are responding to cues for environmental stability and long term resource availability. With good environmental conditions, the optimal life history is to delay child bearing and to invest time in finding a mate. In contrast in the Type I pathway, children receive different environmental cues. Parents are fighting with one another, are neglectful, or the father is absent entirely. These parental behaviors transmit a message that environmental conditions are difficult; they are variable, and perhaps resources are scarce. To delay child bearing to find an ideal mate may prove costly because conditions might become so poor as to make child bearing impossible. Rather than a long courtship and a permanent pair bond, greater reproductive success may be obtained by accepting any partner and starting parenthood early. For boys, aggression may help to deter rivals for female affection. For girls, an early puberty could promote a life history of early reproduction.

In Belsky's hypothesis, children about to embark on these different life histories are only 0-7 years old. They clearly lack the cognitive sophistication to weigh differences in current parental behavior as predictors of future environmental conditions and the optimal timing of parenthood. Thus, the process must be an unconscious one. Children would have to be genetically prepared to be sensitive to cues such as parental conflict and father absence. Their bodies and behaviors change in response to these cues because they have been genetically programmed to do so. One assumption is that over evolutionary time, any child who responded to family conflict and father absence with an earlier reproductive effort had, on average, greater reproductive success. To the extent that such responsiveness to environmental cues was a heritable trait, the genes supporting it would have spread throughout the population, making this life history a genetically-prepared one. Another assumption is that during the Pleistocene, childhood environmental conditions were reliable predictors of environmental conditions some 5 to 10 years later, when the children actually enter into reproductive age. If environmental events cycled more rapidly than a 5-10 year period, then environmental conditions during childhood would be unreliable cues about later environments. Hence, responsiveness to these early environmental cues would not yield any evolutionary benefits.

Ellis, McFadyen-Ketchum, Dodge, Pettit, and Bates (1999) noticed that several evolutionary explanations exist for differences in life histories. For example, Belsky et al. partly built their theory on earlier work by Draper and Harpending (1982; 1988). Ellis et al. saw that Draper and Harpending emphasized the availability of men to women as the crucial environmental cue. In an environment in which men are in short supply, women cannot afford a long courtship, and should benefit evolutionarily from short term matings and early child bearing. Ellis et al. pointed out that Belsky et al. maintained father absence as one environmental cue in their theory, but shifted the emphasis of their theory to intra-family conflict and resource undependability. Thus, in their theory, even a two-parent family could promote the Type I life history through these parental behaviors. Furthermore, the effect of external conditions is an indirect one, mediated through parental behaviors. Presumably, a lower class family with an unstable employment history and a father who was only partially present could avoid the Type I life history in their children if their parenting styles were supportive and the parents managed to endure the environmental hardships with good humor and affection. Ellis et al. thus found that these theories, although intellectually close, actually lead to somewhat different predictions about life histories.

Ellis's own position is closer to Draper and Harpending's original hypothesis than Belsky et al's. Ellis does not believe that it necessarily makes adaptive sense to delay reproduction under good environmental conditions. Clearly, good health and better nutrition leads to an earlier menarche, as evidenced by a secular trend towards younger ages of first menarche. Rather, he believes that the "conditional" feature of the early environment is in relationships with men (in terms of father-mother and father-daughter relationships and those with unrelated men). Girls who witness unstable relationships with their fathers or with their mothers' boyfriends come to expect that men form transient relationships and do not invest heavily in the care and provisioning of children. This pattern prepares them for the Type I life history that accelerates the onset of sexual activity and reproduction. Girls whose early family experiences are characterized by harmonious family relationships (high marital stability, provisioning by father) are hypothesized to develop in a manner that slows the onset of sexual activity and reproduction, makes girls more coy about sexual relationships, and facilitates their long term pair-bonding with men.

Such evolutionary theories encounter a good deal of resistance to their acceptance. Some critics oppose any linking of biology and behavior for reasons of ideology and politics rather than science (Lewontin, Kamin, & Rose, 1984). Nonetheless, they also make substantive points about the difficulty of verifying evolutionary explanations of behavior. One non-ideological critic, Maccoby (1991), presented a counter-point to Belsky's life history theory. She argued that under unstable environmental conditions, women could not benefit from more rapid reproduction because of low rates of offsprings' survival. Her view was that the chain of inference in Belsky et al. theory was not tightly reasoned. For example, Belsky et al. performed no mathematical modelling to link environments, life histories, and evolutionary gains and losses to rates of pubertal maturation. Finally, the data linking these family environmental cues to early puberty and the other life history traits come from causally ambiguous correlational research designs. Despite these weaknesses, the evolutionary theories could still be correct, if not in their exact interpretation of the phenomenon of life histories, then in their identification of different developmental life histories. In the next section of this chapter, I give the behavioral genetic critique of this life history evidence. Once the difference between the behavioral genetic and evolutionary perspectives is understood, data from the *National Longitudinal Study of Adolescent Health* are used to evaluate both perspectives.

1.2   The Behavioral Genetic Critique of Evolutionary Life Histories

Another source of variation in the onset and speed of pubertal development is genetic variation. Twin studies find that monozgotic (MZ) twins are more closely matched in pubertal development than dizygotic (DZ) twins (Pickles et al., 1998; Fishbein, 1977; Kaprio et al., 1995; Treloar & Martin, 1990). These studies all yielded substantial heritabilities for menarcheal timing. In accord with the heritability of age-of-menarche, mother's age of menarche is a strong and direct predictor of daughter's age-of-menarche (Campbell & Udry, 1995).

Treloar and Martin (1990) and Kaprio et al. also detected considerable dominance variation in age-of-menarche. Dominance variance was indicated by a DZ twin correlation that was less than one half the value of the MZ twin correlation. MZ twins share all genetic variance in a trait because they are genetically identical, but DZ twins share 50% of additive genetic variance and 25% of dominance variance because they inherit different genes as well as similar ones. If the genetic trait variation were nothing but genetic dominance variation (a possible but not physiologically plausible situation), then the DZ twin correlation would be ¼ the MZ twin correlation.

Life history theory recognizes that substantial genetic variation exists underlying pubertal timing, but that in addition, it suggests that important environmental variation exists above and beyond the genetic variation. Behavioral genetic research designs also yield evidence on *environmental* components of variation. Two environmental variances are distinguished. Nonshared environmental variance arises from all environmental influences (i.e., anything but genetic make-up) that operate to make family member dissimilar. For example, random embryological events contribute to nonshared environmental variation. If trait variation totally arose from this source, then siblings would correlate zero for the trait. Nonshared influences can also arise when children react idiosyncratically to shared environmental exposures (e.g., divorce). In contrast, shared environmental variance arises from all environmental influences that are correlated across siblings and that also have similar effects on them. Social class is one example of an influence highly correlated between siblings. Thus, when social class differences influence a trait such as vocabulary knowledge, they operate as a shared influence that increases sibling behavioral resemblance.

According to behavioral genetic research, the environmental variance is nonshared for age-of-menarche and pubertal development and timing. In a twin study, shared environmental variation is inferred when the DZ correlation is greater than one half the MZ correlation. More intuitively, consider the case when the MZ and DZ twin correlations are equal in magnitude. Such a pattern would eliminate genetic variation, because the genetically more alike MZ twins are not more alike in the trait. Given common assumptions of the twin method (modest or no assortative mating for a trait, additive genetic variance), the only inference possible is that exposure to shared environments has made both kinds of twins alike. This inference can be more strongly confirmed if data are available from another type of relative: unrelated siblings reared together. Their correlation should equal that of the twin siblings because only shared environmental influences contributes to sibling similarity.

According to the logic of variance components, I would argue that environmental influences highlighted by the evolutionary perspective should operate primarily as shared ones. Father absence or the presence of unrelated males in a household is a shared exposure for siblings (if they are reasonably close in age). The quality of the marriage and quarrels and fights between parents are also likely to be exposures shared by siblings, especially in the critical 0-7 year period. Although parents do treat siblings dissimilarly, parental treatment effects also tend to operate in a shared manner because differences within a family are often smaller than differences between families.

From this argument for shared effects, the components of variation in menarcheal timing are wrong if variation in timing of puberty was due to early environmental cues. DZ twins who are the same age should receive a similar "imprint" for a Type I or Type II developmental pathway; thus they should be alike to the same degree that MZ twins are alike. Even if genetic factors also had some influence on variation in a physical trait, such as menarcheal timing, shared environmental influences could still be informed from the pattern of MZ and DZ correlations. For example, in the case in which the MZ twin correlation equals .70 and the DZ twin correlation equals .50, both genetic (40% of total) and shared environmental (30% of total) variation are large components of total variation, with the remainder being reliable nonshared environmental variation and measurement error (see Rowe, 1994 for details of variance estimates).

The evidence from twin studies, however, indicates that shared environmental factors do not contribute to variation in measures of puberty. Ellis et al. (1999) suggest a methodological reason for the lack of shared environmental influence. They point out that genetic dominance in trait variation can produce underestimations of shared environmental effects in twin studies. Although this argument has merit and should be examined empirically, another possibility is that the variance components are correct and that the life history theory of puberty is false. The next section describes an alternative to the life history view.

1.2.1   The Selection Hypothesis for Variation in Life History Traits

Why do the daughters of emotionally warmer parents have a more delayed puberty (Ellis et al., 1999). Why do the daughters of single mothers reach menarche earlier than do daughters living with two parents (Moffitt et al., 1992)?   An alternative explanation of these findings, and of other surprising results generated by studies testing life history theory, is that these associations may arise from *selection* effects.

As described in research methods textbooks, selection is a methodological confound.   In an experimental study, people are assigned randomly to the treatment and control conditions so that the two groups can be equated for unmeasured variables that may affect experimental outcome.   For example, if (unaware to the experimenter) eye color were an influence on behavior, randomization would nonetheless equate the treatment and control group because the proportions of subjects with blue, brown, etc. eye colors would be about the same in both experimental condition (given a reasonable sample size).   No process of random assignment puts women into the groups of married vs. single parent, or warm and loving vs. harsh and inconsistent in discipline.   Because social groups are not created by random assignment, they may be unequal for variables that influence behavioral development. One variable that may be unequal across social groups is the distribution of genes that influence certain behaviors and traits.

Thus, the selection process permits another interpretation for the early maturation of girls who experience family stressors such as conflict with a parent or step-parent: Their mothers may have been early maturers who transmitted a predisposition towards early physical maturation to them genetically.   In one version of a selection hypothesis, these early maturers may have sex earlier and may marry men who are more impulsive and lower in IQ. The kind of men they marry may lead to less provisioning by fathers and to a greater likelihood of divorce. Furthermore, to the extent that difficult personality traits are heritable, their offsping may possess traits that intensify family conflicts. Although Moffitt et al. (1992) do not offer this exact model, they believed that a "genetic transmission" model offers a more parsimonious explanation of their findings than life history theory.

Although highly speculative, the genes   that affect daughters' puberty might also produce different heritable traits in the mothers, including traits that affect   parenting behaviors. Although parenting variables are usually thought of as "environmental influence," variation among parents in their treatment of children, being nothing more than the behavior of parents, also has a genetic component (Plomin & Bergeman, 1991; Rowe, 1994). To cite one example from this extensive literature, Losoya et al. (1997) compared the parental treatment sibling correlations for adult MZ twins, adult DZ twins, and adult adoptive unrelated siblings.   The adult siblings all had children under 9 years of age. These correlations rank ordered as expected from genetic theory, with the adult MZ twins most like, the adult DZ twins less alike than MZ twins, and the adult adoptive siblings no more alike than randomly paired individuals, despite sharing their family histories. The variance components showed that parenting was heritable, with no shared environmental influence but substantial nonshared environmental influence. Thus, if the same genes affect both menarcheal development of daughters and maternal parenting behaviors, the relationship between parenting behavior and puberty may be spurious (non-causal).

Nevertheless, Ellis and Graber (in press) used their data to defend the life history theory. Girl's pubertal maturation was negatively correlated with maternal depression, marital conflict, and father absence. However in support of a genetic transmission model, there was also a positive relationship between mother's age-of--first birth and daughter's pubertal development. Nonetheless, once mother's age-of-first birth was statistically controlled, the relation of the life history variables (i.e., marital conflict, poor family relationships, father absence) to the outcome of early maturation was only slightly reduced, supporting life history theory. Other findings also favored a life history interpretation. For example, the effects of parental depression on pubertal development appeared to be mediated through family interactions (Ellis and Graber, in press).

These data are not totally decisive, however. Ellis and Graber's first test, statistically controlling the mother's age-of-first birth, was imperfect. It would be better to control her age of menarche, which was unavailable information. Any unreliability in the covariate would make the statistical control result a "low ball" estimate of the true effect of controlling the right covariates. Their complex pattern of findings could be the result of selection, albeit to construct an argument of how the selection produces these particular outcomes would be a difficult task.

In the remainder of this chapter, I examine data from the sibling pairs sample in the *National Longitudinal Study of Adolescent Health* (Add Health) to compare the "genetic transmission" and "life history" theories of pubertal maturation and its correlates. The first question investigates the magnitude of shared environmental variation in pubertal maturation in females. The Add Health research design improves on the twin studies cited above because non-twin kinships were included. The second question concerns the replication with different but comparable measures, the findings of life history correlates of pubertal maturation. The third question is what to make of this covariation; is it genetic, shared environmental, or nonshared environmental?

At the outset, some advantages and limitations of the Add Health genetic study should be noted. On the plus side, it is a fully informative sibling design that included twins, full siblings, half-siblings, and unrelated children in the same household. It can yield estimates of genetic and environmental effects. On the minus side, age-of-menarche was only recorded to the nearest year. For this reason, this measure is imprecise. Furthermore, the data on family environment are concurrent. Thus, the design cannot test the prospective influence of the family environment in childhood on pubertal growth and development in adolescence. Thus, the design assumes stability in family conflict, father absence, etc. such that the conditions observed in adolescence extended back into childhood, and/or it tests a modification of life history theory (i.e., sensitivity to current environments). The sample was heterogeneous, including a number of racial and ethnic groups that vary in rates of maturation, which complicates the analyses. Finally, the teenager was the respondent for all measures in the analyses here. With these caveats, I turn to a description of the sample and presentation of findings.

2.　　SAMPLE AND PROCEDURE

2.1　　The Data

The National Longitudinal Study of Adolescent Health (Add Health) was designed to assess the health status of adolescents and explore the causes of adolescent health-related behaviors. Add Health began with a total sample of over 90,000 adolescents surveyed in school. The primary sampling frame was all high schools in the United States that had an 11th grade and had an enrollment of at least 30 students. A random sample of eighty high schools was selected from this sampling frame, taking into consideration enrollment size, region, school type, ethnicity, and urbanicity. The largest feeder school for each high school was also included in the sample. Seventy-nine percent of the schools initially contacted agreed to participate. Schools that refused to participate were replaced by another school in the same sampling stratum, resulting in a final sample of 134 schools. Within these schools, 90,118 of the 119,233 eligible students (75.6%) in grades 7 to 12 completed a self-administered instrument for optical scanning.

Using both school roster information and information provided by adolescents during the school interview, a random sample of 15,243 adolescents was also selected for a detailed home interview. This sample was stratified by gender and age. The in-home interview was completed by 12,188 (79.5%) of these adolescents. In addition to the core sample, a number of sub-samples was also selected for the home interview. They included samples of disabled adolescents, adolescents from well-educated African American families, adolescents from typically understudied racial and ethnic groups, and a special sibling pairs sample. Overall, 20,745 adolescents completed the in-home interview. Details on the Add Health Study have been reported elsewhere (Resnick et al., 1997).

Included in the 20,745 adolescents is the Add Health *pairs sample* (N = 3,139 sibling pairs), which was selected using information from the in-school questionnaires and school rosters. Specifically, all adolescents who were identified as twin pairs, half-siblings, or unrelated siblings raised together were selected for the home interview. Home interview data were obtained from both the target adolescent and his/her sibling. The sibling pairs were selected regardless of whether they were present on the day of the school interview, and regardless of whether the siblings attended the same school. A probability sample of full siblings was also drawn. The unrelated sibling pairs also included cousins who resided in the same household at the time of the survey. Thus, the pairs sample consists of all genetically informative sibling pairs. The average age in both the full sample and the sibling pairs sample is approximately 16 years (SD = 1.7).

The pairs sample contains 204 opposite-sex DZ twins. The majority of the same-sex twins were diagnosed as either MZ (N=247) or DZ (N=200) on the basis of their self-reports of confusability of appearance. Confusability of appearance scales has been found to have greater than .90 agreement with zygosity determination on the basis of DNA genetic markers (Spitz et al., 1996). Eighty-nine twin pairs of uncertain diagnosis were further classified by matches on genetic markers (47 DZ, 42 MZ). Twins were diagnosed as MZ if they were the same for 5 or more genetic markers (error rate of about 4/1,000 or less) and DZ if they were different at one or more markers. Zygosity could not be ascertained for an additional 43 twin pairs. For the present analyses, these undecided pairs (UD) were omitted. Unrelated siblings living together for less than one year were also removed from the sample. Although the majority of households had just two siblings enrolled in the study, in households with multiple siblings, all possible pairs were made (e.g., siblings A, B, and C form pairs AB, AC, and BC). The use of all pairs makes the significance tests reported later slightly liberal.

The data reported here are from the in-home Wave I questionnaire. This in-home interview was completed between April and December 1995. All respondents were given the same interview, which took form one to two hours to complete. All data were recorded on lap-top computers. Sensitive questions were asked via audio files drawn from the hard disk to coincide with presented questions. In addition to maintaining the data security, this minimized the potential for interviewer bias.

To reduce the heterogeneity of the sample, this study focuses on only female-female sibling pairs in three ethnic groups: African Americans (N=334), Whites (N=836), and Hispanics (N=114; primarily Mexican-Americans). These three ethnic groups form the majority of the sibling pairs sample. The sample sizes in number of sibling pairs were: MZ twins, 109; DZ, 98; full sibling, 289; half-siblings, 87; and unrelated siblings, 59. The total sample size is 1,284 adolescents. In the analyses of menarcheal age, girls who had not experienced menarche (13% of girls) were omitted.

2.2 Measures

Girls were asked two questions about pubertal maturation, one on the developmental of their breasts and the other on the development of their body curves. Both breast and body curve development were measured on 1 to 5 response formats, with one being "about the same as in grade school" and 5 being "a whole lot bigger or curvaceous than in grade school." A 5 point answer to the breast development item also included the statement, "...they are as developed as a grown woman's breasts." The two items were summed to form a scale score, which was later adjusted for its association with age and ethnicity, as described below.

A single item requested the girl's age of menarche to the nearest integer age. Age and ethnicity/race adjustments to menarcheal age are described below.

Adolescent family warmth was assessed using a 5-item scale (alpha = .77). Adolescents reported how much they agreed with the following statements: "people in your family understand you," "your family pays attention to you," "your family has fun together," "your parents care about you," and "you feel like you want to leave home" (item reversed). Responses ranged from 1 = "not at all" to 5 = "very much." Items were averaged to create a single scale score, with higher scores representing greater levels of family warmth. The mean score from the entire Add Health sample was 3.96 (SD=0.70).

Time together was a two-item scale that summed responses to "time spent with siblings" and "time spent with friends." This scale was used in the evaluation of the "equal environments assumption" of the twin and sibling method, as described below. The time together reports of sibling A and B were averaged.

3. RESULTS

3.1 Age and Race

Two variables create possible complexities for the analysis of pubertal maturation: age and ethnic/racial group. Age-of-menarche correlated positively with age, most strongly for the African Americans (r=.31, p < .001; Hispanics, r=.25, p < .05; Whites, r=.26, p < .001; non-menarcheal subjects omitted).   One interpretation of these findings is some secular change has occurred. This interpretation, however, is implausible for such a small age range. Alternatively, our interpretation is that girls "telescope" their reports of menarche to be closer to their current age (see Pickles et al., 1998). Thus, an older girl gives a more biased report of her menarcheal age because she falsely recollected her menarche as closer to her own age than it actually was. This telescoping bias may have been increased by the limits on the reporting; girls gave their age-of-menarche to the nearest age, so that older girls also may have made more rounding errors.

The pubertal maturation scale also correlated with age. This correlation does not represent a bias, however, because older girls necessarily are more mature. Nonetheless, it is a bias for a sibling analysis, because the age effect contributes more to the similarity of age matched MZ or DZ twins than to age discrepant non-twin sibling pairs.

Ethnic differences were also investigated. To illustrate this telescoping bias for menarcheal age, we restricted the sample to girls less than 15 ½ years of age. African American girls matured earlier than White girls (mean African American girls, 11.6 years, SD=1.1; mean White girls, 11.9 years, p < .001). The younger Hispanic girls also matured earlier (M=11.2 years), but their small number (N=17) suggests caution in interpreting this finding. Because sibling A and B were always the same race, racial differences would tend to inflate the sibling correlation for age-of-menarche, if adjustment were not made for the racial/ethnic differences.

Thus, for the analyses that follow, adjusted scores for all the girls were used.   Age-of-menarche was regressed on age, two variables representing race (Hispanic=1 vs. all others=0 and African American=1 vs. all others=0), and the interaction of the second race variable with age.   The interaction term was added to allow for the greater telescoping effect in the African American subjects.   The residuals from this analysis were used in the later analyses;   they represent menarcheal age with the telescoping and racial/ethnic differences removed.   On this adjusted variable rescaled back into the original units, the majority of girls reported menarche between ages 10 and 14 (10 years, 7.4%, 11 years, 18.8%, 12 years, 32.1%, 13 years, 25.6%, and 14 years, 10.7%). The youngest menarcheal age reported was 8 years (.3%) and the oldest was 16 years (also .3%). Two percent reported menarche at age 9 years and 2.8% reported it at 15 years.

Another regression equation was used to adjust the pubertal maturation scores. This regression equation included age and the two race variables, but omitted any interaction term. It is noteworthy that the direction of the racial effect was different than for menarche, as the African American girls reported the lowest mean level of maturity for breast and body curves development. The residuals from this regression equation were used in the later analyses. They represent the timing of pubertal maturation. A positive residual means that a girl was more advanced in pubertal maturation than other girls of her age and race; a negative residual has the opposite interpretation.

3.2   Genetic and Environmental Influences on Pubertal Development

Table 1 presents the sibling correlations (intraclass correlations computed via the double-entry method) for menarcheal age and timing of puberty. The sample sizes were smaller for the age-of-menarche analyses because non-menarcheal girls were excluded. For both variables, the correlations indicated genetic influence on variation. The MZ twins were most alike in pubertal growth, whereas the unrelated siblings were least alike.   The ordering of the correlations for pubertal timing closely followed the degree of genetic relatedness: 1.0 for MZ twins, .50 for DZ twins and full siblings, .25 for half-sibling, 0 for unrelated siblings. For menarcheal age, the half-siblings were more alike than expected by an additive genetic model, but this inconsistency could reflect sampling variation.

The DeFries-Fulker (DF) regression method yields estimates of shared environmental effects (c^2) and heritabilities (h^2) using all the sibling groups in one analysis (DeFries & Fulker, 1985).   The estimates for age of menarche were: $h^2 = .44$, $\sigma_e = .15$, $p < .001$; $c^2 = .08$, $\sigma_e = .09$, n.s.. The heritability of pubertal timing was close in magnitude to that for menarcheal age ($h^2 = .40$, $\sigma_e = .15$, $p < .001$). Its estimate of shared environmental effect was negligible ($c^2 = .00$, $\sigma_e = .08$, n. s.). These results conflicted with the life history theory of pubertal growth as interpreted by behavioral genetics because the majority of environmental variation was nonshared variation, not shared variation. Assuming an approximate reliability of .80 for these measures, the nonshared environmental variation would be .28 for menarcheal age and .40 for timing of puberty.

Table 1. Sibling Correlations for Menarcheal Age and Pubertal Timing

Sibling Group	Menarche		Pubertal	Timing
	r	N pairs	r	N pairs
MZ Twins	.47	88	.39	106
DZ Pairs	.34	77	.23	94
Full Siblings	.30	232	.21	281
Half Siblings	.30	66	.09	86
Unrelated	-.06	42	-.03	56

The sibling analysis assumes that siblings who spend more time together are not more alike in pubertal maturation than siblings who have less social contact with one another. That is, pubertal maturation should not be moderated by the social influence of one twin on the other.

To evaluate this "equal environments" assumption of the twin method, a regression equation was computed within the *full sibling* group. The equation needs to be computed on the full sibling group to control for the genetic relatedness (R) of sibling pairs (a technical note: while full siblings do differ around a mean of 0.50 in genetic relatedness, this variation is slight). That is, by holding genetic relatedness constant, we can determine the independent influence of time together on siblings' behavioral resemblance. The pubertal growth of sibling A was regressed on that of sibling B, time together, and the interaction term, which was the product of sibling B's pubertal growth and time together. A positive and significant interaction term would indicate that siblings who had greater social contact became more alike in their pubertal maturation. If true, then the greater similarity of MZ twins than other sibling types could be a consequence of the MZ twins' greater social contact. Neither the regression equation for menarcheal age nor the one for pubertal timing yielded a significant interaction effect. This analysis was repeated with the siblings' age difference in years replacing time together as the moderator variable. The within-pair age difference (absolute score) also failed to moderate the full siblings' similarity for pacing of pubertal development. Thus, the assumption was upheld.

3.3   The Relation of Pubertal Development to Family Structure

The second research question concerned the association of pubertal growth with family structure. Three levels of family structure were distinguished: two parent, single-parent (always mother only), and mother living with a step-father family. Of the adolescents in these three living arrangements, fifty-eight percent lived with two parents, 28 percent with mother only, and 14 percent lived with a step-father. Siblings who disagreed about their family type or who lived in some other family arrangement (i.e. step-mother) were omitted from these analyses (N=298). Both mother and stepfather and mother-only homes should increase girls' exposure to unrelated males, testing one form of the evolutionary hypothesis. It is important to emphasize again, however, that this evolutionary family structure variable codes the status of the adolescents' living arrangement at the time of their interview, not the kind of environment to which they were exposed during early childhood.

To determine whether an association existed, one-way analyses of variance (ANOVA) were conducted separately by race, with puberty as the dependent variable and the three level family structure variable as the independent variable. The result was that the ANOVA for menarcheal age was statistically significant only for African Americans. In this group, the ANOVA explained 5% of the total variance. The direction of the effect was wrong, however, was wrong to support the life history hypothesis. The means ordered from the earliest menarcheal development in the *two parent* families (M=11.7 years, SD=1.3 years) to later and about equal ages of menarche in the mother only (M=12.3 years, SD=1.2 years) and step-father families (M=12.3 years, SD = 1.2 years). The direction of the relation was as predicted for Whites, with the most delayed menarcheal development in the two parent families, but the group differences were statistically non-significant, despite a good sample size.

For pubertal timing, the pattern of means favored the evolutionary hypothesis in the Hispanic and White groups. With these groups combined, the family mean differences just attained statistical significance (F = 2.95, df = 2,713, p < .05). The means were: two parent families, M = -.10, SD=.90, mother-only families, M = .09, SD=.85, and step-father families, M=.07, SD=1.00. The effect size from two parent to mother-only families is about 2/10ths of a standard deviation. This analysis supported a prediction of evolutionary life history theory that puberty is more advanced for children in other than two parents families.

3.4   Parental Warmth and Pubertal Development

Parental warmth was correlated with later pubertal development (r=-.13, p < .05), but not significantly correlated with menarcheal age (r=.02). Within racial groups, the association between warmth and puberty was only statistically significant for Whites (r=-.17, p < .001).

We also tested for mediation of the family structure – pubertal timing association by parental warmth. In whites, the mean level of warmth differed among the three family structure groups (two parents, M=4.0, SD=.67, mother only, M=3.7, SD=.73, step-father, M=3.8, SD=.74; F = 7.81, df = 1, 630, p < .001). In a multiple regression analysis, pubertal timing was predicted from both warmth and family status (where two parent = 0; mother-only = 1; step-father = 1). In this regression equation, only the term for warmth was significant, both on the full sample and on the White group only. Thus, warmth mediates the association between family type and puberty (although both may be influenced by unmeasured third variables).

An analysis that decomposes covariation into genetic and environmental components is somewhat stymied by the weak associations of the evolutionary life history variables with pubertal growth. To decompose an association explaining only 3% of the variance to genetic and environmental components is difficult, unless sample sizes are very large, because of sampling variation. Nonetheless, there are two distinct views of this association. One is that it arises from genetic influences common to both warmth and pubertal development; this is the "genetic transmission" hypothesis, because warmth in a parent could be affected by the same genes that influence pubertal timing in the children. A possible second hypothesis, that the association arose from shared environmental influences common to both variables, is rendered unlikely by the result for puberty above. No shared environmental variation was found in pubertal timing ($c^2 = .00$).

Table 2. Cross-Trait, Cross-Sibling Correlations for Warmth and Pubertal Timing

Group	r	N	P
MZ twins	-.08	136	.37
DZ twins	-.24	110	.01
Full siblings	-.06	404	.20
Half-siblings	-.00	81	.96
Unrelated siblings	-.05	89	.62

Table 2 displays the cross-trait, cross-sibling correlations between warmth and pubertal timing for White adolescents. These correlations represent the essence of a genetic transmission hypothesis. Specifically, if genetic effects are common to warmth and pubertal timing, then the cross-trait, cross sibling correlations should rank order by the coefficient of genetic relation (R) of the sibling pairs. This expectation was not met, however. In MZ twins, warmth in twin A correlated -.08 with pubertal timing in twin B. This was about the same magnitude as the same cross-trait, cross-sibling correlation in the unrelated siblings (r = -.05). MZ twins are more genetically related (R= 1.0) than unrelated siblings (R=0), but hardly any more cross-trait correlation existed. The only statistically significant correlation was the one found for DZ twins (r = -.24). Their relatedness (R=0.5) is the same magnitude as for full siblings, but the latter group produced only a weak cross-trait, cross-sibling correlation (r = -.06). Although these results fail to support a genetic transmission hypothesis, caution is warranted because the small magnitude of the original within person correlation(r = -.17) would make detecting any systematic pattern with these sample sizes extremely difficult.

4.   DISCUSSION

In this chapter, the evolutionary life history theory of pubertal timing has met a behavioral genetic counter-response. If this were a boxing match, this clash of viewpoints has not left a clear winner. The behavioral genetic view gave no knock out punches, but enough punches were thrown to put the evolutionary life history theory off balance. The evolutionary life history theory also threw a few hard punches. The reader may declare a winner.

One strong punch from the behavioral genetic view was in the results from the DF analyses of both menarcheal age and pubertal timing. In both analyses, about 40% of total variation was explained by genetic variation. Moreover, almost none of the environmental variation was due to shared environmental influences. This latter finding was a blow to the life history theory because conditions such as exposure to a step-father or parental coldness should induce precocious puberty in all daughters in a family, and therefore, be "read" in a behavioral genetic research design as a shared environmental effect. The stronger this environmental imprinting, the greater the amount of shared variation that should be found in the total variation. Its near total absence is problem for evolutionary life history theory, as noted by Ellis et al. (1999). Furthermore, many personality traits are heritable, and the choice of life history strategies may be influenced by genetic inclinations towards high-investment parenting behaviors (MacDonald, 1997) rather than by the specific genetic influences on puberty.

The evolutionary view counter-punched with the findings on the covariation between warmth and pubertal timing. If genetic transmission accounted for the co-occurrence of warmth and pubertal delay in families, then the behavioral genetic decomposition of variance should have found that common genes mediate the correlation between warmth and puberty. The test of this hypothesis, the magnitude of the cross-trait, cross-sibling correlations (i.e., warmth in sibling A with pubertal timing in sibling B), found little evidence of a genetic effect.

The failure to find genetic effects on the covariation between warmth and pubertal timing makes it less likely that mean differences between groups are due to genetic differences. For example, another finding discussed was that the daughters in mother-only and mother with step-father families (at least for Whites and Hispanics) had more precocious puberty than those daughters in two parent families. The genetic selection argument is that the mothers of the daughters in the latter two family types had the genes for earlier puberty (and themselves had more precocious development), which were then passed genetically to the daughters The present study did not test this hypothesis directly. However, warmth also varied among the family structures, and accounted (in a regression) for the relation of family structure and pubertal timing. Thus, we might have expected a genetic correlation between warmth and pubertal timing if the mother-only and step-father daughters differed genetically from their counterparts in two parent families. Of course, it is still possible that the genes that affect puberty do not affect warmth but still differed among the family structures, making the association between family structure and pubertal timing spurious. Clearly, a more direct test than afforded by this study of the effect of the mother's developmental history is needed.

The difficulty in tracking down the source of these life history associations with puberty may be attributed to one the small effect sizes of the life history variables. Small effect sizes can lead to complex findings because associations become unreliable in small samples. The failure to replicate the findings precisely across the different ethnic and racial groups admits to this "small effect size malaise."   For example, in African Americans, the most menarcheally precocious girls were those in two parent families, the kind of family that should be protective against precocious puberty.   This kind of family was less common (32%) than mother only families (54%). In White girls, the situation was reversed (68% in 2-parent families and 18% in single mother families). Nevertheless there is no easy explanation for why either a genetic selection or evolutionary process should make the daughters of two parent families mature earlier. Slightly different problems occurred for the pubertal timing variable.   Here again, ethnic differences existed, with the relation between warmth and pubertal timing stronger in White girls than in African American or Hispanic girls. Too, the percentage of variance explained was small; in White girls, just 2.9%.

So the boxing event between the evolutionary life history and behavioral genetic views must be repeated by studies using other analytic approaches. These new rounds may decide the issue. High on the list of methodological priorities is a collection of a complete history of the mother's menarcheal development.   A research design that includes adoptive unrelated parent – child pairs and biologically related parent child pairs would provide a more direct evidence of genetic transmission in the family and of any genetic mediation of the association between parental warmth and children's puberty.   In molecular genetics, a discovery of genes that influence the timing of puberty would also greatly enhance the research design possibilities because children's genetic dispositions would become a measured variable in the research design.   The questions about physical development are too fascinating to be left without satisfactory answers.

ACKNOWLEDGEMENTS

This research is based on data from the National Longitudinal Study of Adolescent Health, a program project designed by J. Richard Udry (PI) and Peter Bearman, and funded in part by grant P01-HD31921 from the National Institute of Child Health and Human Development to the Carolina Population Center, University of North Carolina at Chapel Hill, with cooperative funding participation by the National Cancer Institute; the National Institute of Alcohol Abuse and Alcoholism; the National Institute on Deafness and Other Communication Disorders; the National Institute of Drug Abuse; the National Institute of General Medical Sciences; the National Institute of Mental Health; the National Institute of Nursing Research; the Office of AIDS Research, NIH; the Office of Behavior and Social Science Research, NIH; the Office of the Director, NIH; the Office of Research on Women's Health, NIH; the Office of Population Affairs, HHS; the National Center for Health Statistics, Centers for Disease Control and Prevention, HHS; the Office of Minority Health, Centers for Disease Control and Prevention, HHS; the Office of Minority Health, Office of the Assistant Secretary for Health, HHS; the Office of the Assistant Secretary for Planning and Evaluation, HHS; and the National Science Foundation. Persons interested in obtaining data files from The National Longitudinal Study of Adolescent Health should contact the Carolina Population Center, 123 West Franklin Street, Chapel Hill, NC 27516-3997. Further support for this research came from Grant R01-HD2-1973 from the National Institutes of Health.

REFERENCES

Belsky, J., & Draper, P. (1990). Personality development in evolutionary perspective. *Journal of Personality, 58*, 141-161.

Belsky, J., Steinberg, L., & Draper, P. (1991). Childhood experience, interpersonal development, and reproductive strategy: An evolutionary theory of socialization. *Child Development , 62*, 682-685.

Campbell, B.C., Udry J.R. (1995). Stress and age at menarche of mothers and daughters. *Journal of Biosocial Science, 27*, 127-134.

Calabrese, L.H., Kirkendall, D.T., Floyd, M., Rapoport, S., Williams, G.W., & Weiker, G.F., Bergfeld, J.A. (1983). Menstrual abnormalities, nutritional patterns, and body composition in female classical ballet dancers. *Physician and Sports Medicine , 11*, 86-98.

DeFries, J.C., Fulker, D.W. (1985). Multiple regression analysis of twin data. *Behavior Genetics , 15*, 467-473.

Draper, P., & Harpending, H. (1982). Father absence and reproductive strategy: An evolutionary perspective. *Journal of Anthropological Research , 38*, 255-273.

Draper, P., & Harpending, H. (1988). A sociobiological perspective on the development of human reproductive strategies." In MacDlonald, K. B. (Ed.) *Sociobiological Perspectives on Human Development.* New York: Springer-Verlag, 1988.

Ellis, B.J., McFadyen-ketchum, S., Dodge, K.A., Pettit, G.S., Bates, J.E. (1999). Quality of early family relationships and individual differences in the timing of pubertal maturation in girls: A longitudinal test of an evolutionary model. *Journal of Personality and Social Psychology, 77,* 387-401.

Ellis, BJ, Graber, J. (1998, in press). Psychological antecedents of pubertal maturation in girls: Parental psychopathology, stepfather presence, and family and marital stress. *Child Development.*

Fischbein, S. (1977). Onset of puberty in MZ and DZ twins. Acta Geneticae Medicae et Gemellologiae 26, 151-157.

Graber, JA, Brooks-Gunn, J, & Warren, MP. (1995). The antecedents of menarcheal age: Heredity, family environment, and stressful life events. *Child Development , 66,* 346-359.

Harris, J. (1998). *The Nurture Assumption.* New York: The Free Press.

Kaprio, J., Rimpela, A., Winter, T., Viken, R.J., Rimpela, M., Rose R.J. (1995). Common genetic influence on BMI and age at menarche. *Human Biology, 67,* 739-753.

Lewontin, R.C., Kamin, L.J., & Rose, S. (1984). *Not in Our Genes.* New York, Pantheon.

Losoya, S.H., Callor, S., Rowe, D.C., & Goldsmith, H.H. (1997). The origins of familial similarity in parenting: A study of twins and adoptive siblings. *Developmental Psychology, 33,* 1012-1024.

MacDonald, K. (1997). Life history theory and human reproductive behavior: environmental/ contextual influences and heritable variation. *Human Nature , 8,*327-359.

Maccoby, E.E. (1991). Different reproductive strategies in males and females. *Child Development , 62,* 676-681.

Moffitt, TE, Caspi, A, Belsky, J., & Silva, PA. (1992). Childhood experience and onset of menarche: A test of a sociobiological model. *Child Development , 63,* 47-58.

Pickles, A, Pickering, K, Simonoff, E, Silverg, J, Meyer, J, & Maes, H. (1998). Genetic "clocks" and "soft" events: A twin model for pubertal development and other recalled sequences of developmental milestones, transitions, or ages at onset. *Behavior Genetics , 28,* 243-253.

Plomin, R. & Bergeman, C.S. (1991). The nature of nurture: Genetic influence on "environmental" measures. *Behavioral and Brain Sciences, 14,* 373-427.

Rowe, David C. (1994). *The limits of family influence: Genes, experience, and behavior.* New York: Guilford.

Udry, J.R., & Bearman, P.S. (1998). New methods for research on adolescent sexual behavior. In R. Jessor et. al (Eds.) *New perspectives on adolescent risk behavior,* New York: Cambridge University Press.

Steinberg, L.. (1988). Reciprocal relation between parent-child distance and pubertal maturation. *Developmental Psychology , 24,* 122-128.

Treloar, S.A. & Martin, N.G. (1990). Nonadditive genetic variance detected in a large twin sample. *American Journal of Human Genetics , 47,* 137-148.

Chapter 8

BEHAVIOR GENETIC MODELING OF MENARCHE IN U.S. FEMALES

Debby Doughty and Joseph Lee Rodgers

Key words: menarche, DF analysis, behavior genetic analysis

Abstract: Most previous research has been logically unable to disentangle the genetic and environmental influences on age at menarche. We present data on 1338 kinship pairs from the National Longitudinal Survey of Youth in a behavior genetic analysis, partitioning variability in menarcheal age into genetic and environmental sources. About half the variability in menarcheal age was related to genetic influences, $h^2 = .54$, and almost half to nonshared environmental influences plus error. No influence of the shared environment was found. Motivated by the evolutionary theory of Belsky, Steinberg, and Draper, the influence of family composition / stability was tested as a nonshared environmental influence. In line with previous findings, father absence was associated with a younger age at menarche. Residing with two parents under extreme living conditions may delay age at menarche. No association of family size, birth order, personality, income, or parental education with age at menarche was found.

1.   INTRODUCTION

The age of onset of pubertal development has been linked to biological and psychosocial influences. However, the simultaneous and relative influences of biological and environmental factors on individual differences in development are seldom analyzed in research designs. In this paper, we present data from the National Longitudinal Survey of Youth (NLSY) and conduct a behavior genetic analysis, partitioning variability in menarcheal age into genetic and environmental sources.

1.1   Background

The consequences of early menarche may include an earlier age of first intercourse (Zabin, Smith, Hirsch, & Hardy, 1986; Presser, 1978), marriage (Buck & Stavraky, 1967; Phinney, Jensen, Olsen, & Cundick, 1990; Udry, 1979), and first birth (Wasserman, Rauh, Brunelli, Garcia-Castro, & Necos, 1990; Udry 1979), as well as having more children in the adult years than do women with later maturation (Udry & Cliquet, 1982). Many of these consequences have in part been attributed to "the secular trend" for earlier ages of pubertal maturation. The secular or generational trend describes the period from around 1850 to 1950, during which body size increased along with a concurrent decrease in age of menarche of approximately four months per decade in both poor and affluent children in every country studied. Average age at menarche has changed from 16 or more years a century ago to13 years or less today (Eveleth & Tanner, 1990; Garn, 1980).

Although the consequences of age at menarche are well documented, the biological and environmental precursors of menarcheal timing are less well understood. This paper will examine the role of home environment and heredity on age at menarche.

1.1.1   Home Environment

Several recent studies have examined the effect of family environment on menarcheal timing. These studies focus on family size, birth order, family stress or conflict, and parental absence (in particular father absence). The family size and birth order studies may be motivated by the following propositions: 1) Larger family sizes may be associated with lower socioeconomic status and living conditions, influencing later born girls to reach menarche at later ages. 2) Pheromones may exist in the human to influence the menstrual cycle (McClintock, 1971), thus the presence of older sisters may influence earlier menarche in younger siblings. 3) Younger siblings have an age of first intercourse earlier than older siblings (Rodgers,

Rowe, & Harris, 1992) which may in part be due to earlier maturational timing.

Varying degrees of support have been found for these propositions. Girls with early menarche have been found to come from smaller families with fewer sisters (Stattin & Magnusson, 1990). Malina, Bouchard, Shoup, Demirjian, & Lariviere (1979) reported that later born girls in large families had later menarche in one study (although the differences between birth order categories are not always significant), while another study found earlier menarche associated with later birth order and larger families. Other researchers have found no effect of the number of siblings, or the number of older or younger sisters or brothers on menarcheal age (Jones, Leeton, McLeod, & Wood, 1972; Zacharias, Rand, & Wurtman, 1976).

The effects of family stress on pubertal maturation are also unclear. The accelerating hypothesis, which states that pubertal maturation is accelerated by parent-child distance, has found some support in the mother-daughter relationship (Steinberg, 1988). However, lower stress has also been associated with accelerated pubertal development in girls and emotional distance between the girl and her parents was associated with a slowing of puberty (Ellis, 1991).

A sociobiological model has been proposed to explain maturational timing in terms of how childhood behavioral and psychological problems and family stress may influence early pubertal maturation. This model is part of a broader evolutionary theory of socialization in which early childhood experiences affect how the girl will later apportion reproductive effort both psychologically and biologically (Belsky, Steinberg, & Draper, 1991). Within this framework, individuals reared in unstable, conflictual environments develop poor attachments and problem behaviors. These girls reach puberty at a younger age and engage in early sexual activities oriented toward short term pair bonds, which result in single parenthood and unstable environments for their children (a quantity strategy). Individuals reared in more positive environments have a later age of pubertal onset and defer sexual activity with the goal of establishing more enduring pair bonds and a greater parental investment (a quality strategy). In a partial test of this model (Moffitt, Caspi, Belsky, & Silva, 1992), family conflict and father absence had direct effects on age at menarche while behavior problems and weight (hypothesized to accelerate age at menarche) did not play a role. It was suggested that these results could reflect a heritable association between early maturation of daughters with that of mothers who may have been more likely to marry and have children at a young age and would be more at risk for family conflict and divorce. Subsequent research partially supported these findings in that family relations and depressive affect were found to be associated with age at menarche, although these girls came from intact families and stressful events were not predictive of age at menarche (Graber, Brooks-Gunn, & Warren, 1995). A later study assessed both the relationship of early childhood stress to menarcheal age and whether the association between father absence and early menarche of the daughter is mediated by the mother's age at menarche (Campbell & Udry, 1995). No support was found for either relationship. Only mother's education was predictive of her daughter's age at menarche, which may have reflected eating habits or lifestyle characteristics related to education.

1.1.2   Heredity

Many researchers have noted in their papers that it is difficult to disentangle genetic effects from environmental ones. Genetic influences on menarcheal age have been assessed by correlations between kin pairs and range from .22 to .54 for mother - daughter pairs (Chern, Gatewood, & Anderson, 1980) and from .25 (Chern, Gatewood, & Anderson, 1980) to .39 (Tanner, 1955) for sister pairs.   Because of the inherent confounding, environmental and genetic influences cannot be separately assessed by these correlations.   A genetic design using MZ and DZ twins suggested that timing of menarche is more similar for MZ ($r=.93$) than for DZ ($r=.62$) twins (Fischbein, 1977) and provided stronger evidence of the genetic contribution to variability in menarcheal timing.   In a study of mothers and daughters using the correlation coefficient between relatives' menarcheal age multiplied by the coefficient of relationship (equal to ½ for mother-daughter and sister-sister relationships), heritability of menarcheal age was estimated as .45 (Chern, Gatewood, & Anderson, 1980).   Kinship analyses of menarcheal seasonality suggest a biological basis for menarche peaks observed in the summer months for U.S. females, with heritability estimated at .62 and no influence of shared family environment (Rodgers & Buster, 1994).

In summary, these findings suggest that menarcheal timing is a process with biological, environmental, and psychological components, although the links have not been well defined.   As the influence of family size, birth order, and family environment appears to be the least well understood, we will consider related variables in analyses using a behavioral genetic framework that accounts for individual differences in menarcheal age as the dependent variable by partitioning this variance into genetic, shared environmental, and unique nonshared environmental events.

2.   METHOD

2.1   Data

The data came from the National Longitudinal Survey of Youth (NLSY), a nationally representative sample including 6283 women who were 14 to 22 years of age when first surveyed in 1979. The respondents were followed yearly and in 1984 and 1985 (when they were 19 to 27 years old) were asked their age of menarche in years as well as the month of menarcheal onset. The NLSY files contain detailed information on family background characteristics, and household composition, that permit construction of kinship links (Rodgers, 1996) which support behavioral genetic analyses partitioning variance in age at menarche into genetic, shared environmental, and nonshared environmental influences.

2.2   DF Analyses

The regression model developed by DeFries and Fulker (1985) allows simultaneous testing of both genetic and environmental influences by using scores from multiple levels of kinship pairs to provide parameter estimates of genetic variation (h^2), shared environmental variation (c^2), and a combination of nonshared environmental variation and measurement error. Assumptions of the model are equal shared environmental influences, trivial assortative mating, and additivity.

Scores for each kinship pair were entered into the following regression model:

$$AM_1 = b_0 + b_1 AM_2 + b_2 R + b_3(AM_2 * R) + e \qquad (1)$$

where AM_1 and AM_2 are the ages at menarche of the two members of the kinship pair, and R is the coefficient of genetic relatedness. (In the kinship linking algorithm for this data set: R=.75 for twins (zygosity unknown), R=.50 for full siblings, R=.375 for undefined full or half siblings, R=.25 for half siblings, and R=.125 for cousins.) Within the assumptions of the model, an unbiased estimate of heritability (h^2) is provided by b_3, an unbiased estimate of shared environmental influences (c^2) is provided by b_1, and e is the residual containing variance due to the nonshared environment and measurement error. As neither member of the kinship pair was selected on age at menarche, the data were double entered so that each member of the kin pair was entered in the equation twice, i.e. the first member of the pair's AM was entered as AM_1 and and second member's AM as AM_2, then the second member's AM was entered as AM_1 and the first member's AM as AM_2. Standard errors and conservative tests of significance were computed by adjusting the sample size back to the number of pairs. The following extension presented by Rodgers, Rowe, and Li (1994) was used to test for nonshared environmental influences with a difference score between members of the kinship pairs for the environmental measure indicated by ENDIF:

$$AM_1 = b_4 + b_5 AM_2 + b_6 R + b_7 (AM_2 * R) + b_8 ENDIF + e \qquad (2)$$

Specific environmental variables that may support or extend current theory related to menarcheal age were selected from the data files to be tested in the DF model, and the difference in the kin pairs' scores on these variables was entered as ENDIF in equation 2. The following variables from the 1979 data closest to age at menarche were used as potential psychosocial influences on pubertal maturation.

1. Beliefs about self: perceived control over life, self esteem. These were assumed to be indirect measures of the home environment.

2. Measures of household and family characteristics: intact family (at age 14, lived with both biological parents vs. lived with no male adult in the household), highest grade completed by father, highest grade completed by mother, any member of the household live in public housing in the past year, any member of the household receiving welfare or public assistance, unemployment rate for labor market of the community resident in, poverty status, net income, family size, number of siblings, and number of siblings older than the respondent. These were assumed to provide information on family structure and quality of the home environment and are considered risk indicators for increased conflict and stress.

3.   RESULTS

The mean age at menarche for the 1338 female kin pairs that were linked in the kinship sample was 12.83 +/- 1.57 years. Reported age at menarche ranged from 8-18 years. For the 6025 females reporting age at menarche in the entire NLSY data set, mean menarcheal age was 12.78 +/- 1.57 years. Reported age at menarche ranged from 0-19 years. The kin pairs are constructed in such a way that the oldest youth in the household is separately linked with each other youth in that household. Thus the oldest household member is disproportionately represented in the sample relative to the other youth. However, comparison of the means and standard deviations for reported age at menarche of the kinship sample to the entire NLSY sample suggests that little bias is present on this measure related to the method of kin pair construction.

We ran all analyses using the full kin pairs data set and also using a smaller data set where the oldest member of the household is represented only once. No differences were found between the methods except where the smaller data set gave sample sizes that were too small.[1] Frequencies for each reported age at menarche for the kin pairs sample we used for the DF analysis compared with those for the entire NLSY data set are shown in Table 1 along with the cumulative percent of each sample that is represented at each menarcheal age.

Table 1. Frequency and Cumulative Percent of Entire NLSY Sample and Kin Pairs Sample at Each Age of Menarche

Age	Entire NLSY Sample		Kin Pairs Sample	
	Cumulative Frequency	Percent	Cumulative Frequency	Percent
8	7	.73	3	.07
9	114	2.62	85	1.77
10	242	6.62	169	5.18
11	668	17.62	460	14.44
12	1613	44.39	1284	40.29
13	1756	73.54	1405	68.57
14	841	87.05	784	84.35
15	445	94.44	458	93.57
16	249	98.57	252	98.64
17	64	99.63	54	99.73
18	21	99.98	14	100.00
19	1	100.00		

The overall NLSY sample had a few reported ages at menarche as early as zero. The cumulative percentages suggest the kin pairs are slightly later maturers than the entire sample of NLSY females and that the few individuals reporting extremely early ages at menarche in the overall sample cannot account for the almost 5% cumulative difference by age 13 in the kinship sample. Thus the effects of variables influencing early maturation may be attenuated somewhat for analyses on the kin pairs sample.

The correlations for age at menarche for the kin pairs are shown in Table 2. There is generally increasing correlation for increasing genetic relatedness, except for half-siblings (who had a relatively small sample size).

Table 2.   Kinship Correlations for Age at Menarche

Relationship	Number of pairs	r	p
full data set	1178	.17	.0001
cousins	44	.13	.4002
half siblings	20	-.39	.0817
undefined full or half siblings	146	.28	.0007
full siblings	940	.26	.0001
twins	28	.38	.0466

When we fit the regression model in equation 1 for the full kin pairs data set (N=1178), shared environmental influences that would explain similarities between the kinship pairs on menarcheal age were not significantly different from zero ($h^2 = .52$, s.e. $= .39$, $c^2 = .01$, s.e. $= .19$). Thus the c^2 estimate was dropped from the model, and the heritability was estimated, resulting in $h^2 = .54$, s.e. $= .08$, p<.01. This suggests that a little over half the variability in menarcheal age for this data set is due to genetic influences and almost half the variability is due to nonshared environmental influences and measurement error. Separate analyses by race gave estimates for blacks (N=282) of $h^2 = .73$, s.e. $= .17$, p<.01, and for whites (N=788) of $h^2 = .42$, s.e. $= .10$, p<.01.

Following are the results of separately adding the difference scores for the environmental measures to the model in equation 2.

3.1   Beliefs about self

Of the sibling difference scores for four measures of perceived control and ten measures of self esteem tested, none were found to be significantly related to age at menarche.   Sample sizes for the variables ranged from 1168-1172, large enough to detect effects with considerable power. These results suggest that locus of control and self esteem do not influence age at menarche nor are they influenced by age at menarche. This finding differs from previous ones suggesting that early maturers have lower self esteem.

3.2   Measures of household and family characteristics

For this analysis, *s*ample sizes for each of the measures ranged from 1004-1178, providing enough power to detect influences of the ENDIF variables. The presence of both biological parents in the household versus no male adult in the household was significantly related to age at menarche (N=1104, $\underline{b}$=.01, $\underline{se}$=.005, $\underline{p}$<.05) providing support for previous findings by Moffit, Caspi, Belsky, & Silva (1992), although the data do not provide information on when the father left the household. Father absence was associated with an accelerated age at menarche of 1.32 months contrasted with 2.7 months found by Moffit et al. Mean age at menarche for girls living with both parents was 12.95 +/- 1.53 years (N=2527) and the mean age at menarche for girls living with the mother only was 12.84 +/- 1.59 (N=618), while the overall sample of Moffit et al. had a mean age of menarche just over 13.

Income, poverty status, public assistance or housing, and the unemployment rate in the community in which the girl resided were not significant predictors of menarcheal age for the overall sample of kin pairs. A separate analysis of girls living with both biological parents and of girls living with the mother only revealed that for those from intact families, living in public housing was positively related to age at menarche and marginally significant   (N=834, $\underline{b}$=-.85, $\underline{se}$=.52, $\underline{p}$=.0511) while no relationship between housing status and age at menarche was found for girls living with the mother only. Table 3 shows the mean age of menarche by parental residence and housing status. The means show that for girls living in public housing with both parents, menarche is delayed by 2.64 months relative to girls living in public housing with the mother only, which is very close to the timing difference found by Moffitt et al.

While Campbell & Udry (1995) found that maternal education was related to menarcheal age, we did not find evidence that maternal education was related to age at menarche. Given the association of education and living conditions, the finding that girls living only with their mothers do not significantly differ in menarcheal age regardless of housing/living conditions differs from the suggested interpretation of Campbell and Udry's data.

Family size, number of siblings, and number of siblings older than the respondent were not significantly related to menarcheal age, giving no support for the proposition that girls from larger families with potentially lower SES and poorer living conditions reach menarche at later ages. The proposition that younger siblings may mature earlier than older siblings also was not supported by these analyses. If pheramonal influences are important for menarcheal onset, then perhaps the presence of older sisters, rather than the number of older siblings are more important. We would expect that siblings closer in age would spend more time together, allowing a greater influence of pheromones on the hormonal pathways involved in menarcheal timing if these exist. We were able to identify which females in the household were older sisters so that the influence of age spacing could be determined. A subset of 349 full sister pairs comprised of the oldest and youngest sisters in the household and 7 twin pairs did not reveal any association of age spacing with age at menarche.

Table 3. Mean Age at Menarche by Parental Residence and Housing Status

Housing	Lives with Both Biological Parents			Lives with Mother Only		
	<u>M</u>	<u>SD</u>	<u>n</u>	<u>M</u>	<u>SD</u>	<u>n</u>
Public	13.17	1.84	96	12.86	1.59	108
Non-Public	12.95	1.52	2220	12.85	1.54	482

4.   DISCUSSION

The results of this behavior genetic analysis confirm that the timing of menarche is strongly driven by genetic processes and also that differences in menarcheal timing are almost equally influenced by nonshared environmental influences for this population. Shared environmental factors among kinship pairs were found to have no influence on menarcheal timing. No support was found for the theory that family size, birth order, or personality variables that measure beliefs about the self have an influence on or are influenced by menarcheal age. In agreement with previous research, father absence was found to be associated with earlier menarche.

Girls in intact families living in public housing showed delayed menarche relative to the overall population. Unfortunately, the influence of relational or personal problems on onset of menarche could not be assessed. The specific environmental variables and processes that contribute to differences in time of menarche between related individuals by their action on overall growth rate and influence on hormonal pathways remain to be identified.

Belsky's theory represents a cumulative conditional probability concept. In his own test of the model (see Moffitt et al., 1992), only the psychosocial factors of father absence and family conflict had an additive influence on menarcheal onset. We consider our study more exploratory as to the nature of relationships between the selected environmental measures and menarcheal age rather than a test of this theory, which would need to take into account the cumulative effect of differentially weighted multiple paths, including the effects of *prepubertal* parent-child relations, behavioral development, and rearing conditions which our data did not provide specific measures for.

Our kin pairs sample matured slightly later than the overall NLSY data set they were taken from, yet slightly earlier than the sample in Moffit et al. That this study also found an effect of father absence on accelerating age at menarche provides further support for the importance of considering the family rearing environment in studies of pubertal development. As our model did not find shared environmental influences to be present, the effect of father absence on differences between siblings' menarcheal timing must be related to nonshared genetic or environmental differences. Previous research has suggested that it is not father absence, but stress associated with the single parent household and divorce that accelerates pubertal maturation. Siblings may be differentially predisposed genetically to respond to environmental stress. We might also speculate that if a two-parent family must rear children in public housing, they may be subject to additional obstacles that would increase the level of stress relative to a single mother needing public housing. Belsky suggested that biological systems may have evolved to accelerate pubertal development when reproductive prospects were limited but to delay development when survival is threatened. Further research should examine the curvilinear nature of stress relationships to demonstrate that moderate stress allows utilization of developmental resources while extreme stress may conserve these resources if this is the case for menarcheal onset.

As the environmental measures available were reported after menarcheal onset, we cannot determine whether father absence effects were present prepubertally. Father absence does suggest however, that the home environment may not have been conflict free even if the father was present before menarcheal onset. Future studies should take into account the amount of contact with a nonresiding father and the quality of the relationship before and after he left. That the reports of age at menarche were made retrospectively could also potentially reduce the validity of our findings. However, the congruence of the results with previous research strengthens support for the results.

Despite the noted limitations, the design of this study was able to demonstrate simultaneously the relative effects of heredity and nonshared environmental influences on menarcheal age, ruling out shared environmental effects. Future research should test for differential parental treatment of siblings as well as nonshared genetic influences which may identify gene - environment correlations that influence menarcheal age and whether a link exists between menarcheal timing, genetically and environmentally influenced behaviors, and which of the two reproductive pathways are ultimately chosen - a quantity or quality strategy.

ACKNOWLEDGEMENTS

This research was supported by NIH Grant #R01-HD21973.

REFERENCES

Belsky, J., Steinberg, L, & Draper, P. (1991). Childhood experience, interpersonal development, and reproductive strategy: An evolutionary theory of socialization. *Child Development, 62*, 647-670.

Buck, C. & Stavraky, K. (1967). The relationship between age at menarche and age at marriage among childbearing women. *Human Biology, 39*, 93-102.

Campbell, B. C., & Udry, J. R. (1995). Stress and age at menarche of mothers and daughters. *Journal of Biosocial Science, 27*, 127-134.

Chern, M. M., Gatewood, L. C., & Anderson, V. E. (1980). The inheritance of menstrual traits. In A. J. Dan, E. A. Graham & C. P. Beecher (Eds.), *The Menstrual Cycle*. New York: Springer Publishing Company.

Ellis, N. B. (1991). An extension of the Steinberg accelerating hypothesis. *Journal of Early Adolescence, 11*, 221-235.

Eveleth, P. B., & Tanner, J. M. (1990). *Worldwide Variation in Human Growth*. Cambridge: Cambridge University Press.

Fischbein, S. (1977). Onset of puberty in MZ and DZ twins. *Acta Genet. Med. Gemollol, 26*, 151-158.

Graber, J. A., Brooks-Gunn, J., & Warren, M. P. (1995). The antecedents of menarcheal age: heredity, family environment, and stressful life events. *Child Development, 66*, 346-359.

Hill, J. P., Holmbeck, G. N., Marlow, L., Green, T. M., & Lynch, M. E., (1985). *Journal of Youth and Adolescence, 14*, 301-316.

Jones, B., Leeton, J., McLeod, I., & Wood, C. (1972). Factors influencing the age of menarche in a lower socio-economic group in Melbourne. *The Medical Journal of Australia, 2*, 533-535.

Malina, R. M., Bouchard, C., Shoup R. F., Demirjian, A., & Lariviere, G. (1979). Age at menarche, family size, and birth order in athletes at the Montreal olympic games, 1976. *Medicine and Science in Sports, 11*, 354-358.

McClintock, M. K. (1971). Menstrual synchrony and suppression. *Nature, 229,* 244-245.

Moffitt, T. E., Caspi, A., Belsky, J., & Silva, P. A. (1992). Childhood experience and the onset of menarche: A test of a sociobiological model. *Child Development, 63,* 47-58.

Phinney, V. G., Jenson, L. C., Olsen, J. A., & Cundick, B. (1990). The relationship between early development and psychosexual behaviors in adolescent females. *Adolescence, 25,* 321-332.

Presser, H. B. (1978). Age at menarche, socio-sexual behavior, and fertility. *Social Biology, 25,* 94-101.

Rodgers, J. L. (1996). NLSY Linking Algorithm. Internal document.

Rodgers, J. L., & Buster, M. (1994). Seasonality of menarche among U. S. females: Correlates and linkages. In K. L. Campbell & J. W. Wood (Eds.), *Human Reproductive Ecology: Interactions of Environment, Fertility, and Behavior.* New York: The New York Academy of Sciences.

Rodgers, J. L., Rowe, D. C., & Harris, D. F. (1992). Sibling differences in adolescent sexual behavior: Inferring process models from family composition patterns. *Journal of Marriage and the Family, 54,* 142-152.

Stattin,H., & Magnusson, D. (1990). *Paths Through Life: Vol. 2: Pubertal Maturation in Female Development.*

Steinberg, L. (1988). Reciprocal relation between parent-child distance and pubertal maturation. *Developmental Psychology, 24,* 122-128.

Tanner, J. M. (1955). *Growth at Adolescence.* Springfield: Charles C. Thomas Publisher.

Udry, J. R. (1979). Age at menarche, at first intercourse, and at first pregnancy. *Journal of Biosocial Science, 11,* 433-441.

Udry, J. R., & Cliquet, R. L. (1982). A cross-cultural examination of the relationship between ages at menarche, marriage, and first birth. *Demography, 19,* 53-63.

Wasserman, G. A., Raub, V. A., Brunelli, S. A., Garcia-Castro, M., & Necos, B. (1990). Psychosocial attributes and life experiences of disadvantaged minority mothers: Age and ethnic variance. *Child Development, 61,* 566-580.

Zabin, L. S., Smith, E. A., Hirsch, M. B., & Hardy, J. B. (1986). Ages of physical maturation and first intercourse in black teenage males and females. *Demography, 23,* 595-606.

Zacharias, L., Rand, W. M., & Wurtman, R. J. (1976). A prospective study of sexual development and growth in American girls: The statistics of menarche. *Obstetrical and Gynecological Survey, 31,* 325-337.

Notes

[1] In the smaller kin pairs sample, there were 852 observations and $h^2 = .39$, s.e. $= .44$, $c^2 = .08$, s.e. $= .21$, $p>.05$. After dropping the c^2 estimate and re-estimating h^2, the heritability was $h^2 = .55$, s.e. $= .10$, $p<.05$. With the smaller sample, 848 observations were available to test the influence of living with both biological parents compared to living with the mother only at age 14, b$=.01$, s.e. $= .007$, p$=.0764$. For both the full kin pairs sample and the smaller kin pairs sample the proportion of the first member of the kin pair living with only the mother compared to living with both parents did not differ. Sample sizes were too small using the smaller kin pairs sample to assess the relationship between housing status, parents lived with, and age at menarche.

Chapter 9

EARLY FAMILY ENVIRONMENT, REPRODUCTIVE STRATEGY, AND CONTRACEPTIVE BEHAVIOR: TESTING A GENETIC HYPOTHESIS

Warren B. Miller and David J. Pasta

Key words: early family environment, reproductive strategy, contraceptive behavior, childbearing, implicit social bonding

Abstract: Recently, Belsky, Steinberg, and Draper (1991) theorized that the amount of stress and discord experienced by children in their early family environment helped shape the timing and nature of their adolescent reproductive behavior. Here we examine that relationship and the affect it may have on contraceptive practice during the young adult period, unifying our approach with a theory of social bonding. We collected data on family background, adolescent development, and current contraceptive behavior from a convenience sample of 178 mostly unmarried couples. We used LISREL to construct a two-sex, constrained model of variable relationships across the three time periods, while simultaneously adjusting for selected social-demographic and personal traits and for major situational factors. We found that levels of both affection and abuse in the respondent's family of origin had multiple effects on both adolescent development and adult contraceptive behavior, even when father absence and mother's age at first birth were controlled. We interpret these findings as suggesting the action of a genetic switch that affects developmental timing. We discuss some possible cellular processes that could underly such a mechanism and the probable adaptive consequences of its effects on bonding. We then reconceptualized our findings in terms of the constructs of implicit and explicit motivation. We conducted a second LISREL analysis. The results indicate that our measure of implicit childbearing motivation, which is based on the adolescent development variables and is presumably in large part non-conscious, is far more important in the determination of pregnancy avoidance motivation than our measures of explicit childbearing motivation, which are self-attributive and mostly conscious. We discuss how these findings might relate to earlier work in which only explicit childbearing motivations were measured.

1.   INITIAL COMMENT

With the possible exception of sexual abstinence, it seems unlikely that any of the behaviors which we consider in our contemporary world to be contraceptive in nature were practiced by humans during the Pleistocene, that epoch of geologic time beginning about two million years ago and ending about ten thousand years ago during which Homo Sapiens emerged. If these behaviors were not present in, and therefore not shaped by, the enviroment of our evolutionary adaptedness, how could genes be related to individual variation in contemporary contraceptive behavior? The answer to that question is a complex one but lies partly, we believe, in the specifics of the theoretical framework that we develop in this chapter and then test with data from a sample of sexually active young adults living in California.

2.   THEORETICAL FRAMEWORK

2.1   The Early Family Environment and Adolescent Sexuality

We were originally stimulated in our efforts to understand the possible biological underpinnings of contraceptive behavior by the work of Belsky, Steinberg, and Draper (1991), who theorized that the child's experience with its early family environment shaped subsequent reproductive behaviors in a way that was designed by evolution to maximize reproductive success. Their fundamental argument, subsequently enlarged upon and elaborated by Belsky (1996, 1997), was that a family context characterized by discord, stress, and resource deficiency tended to produce insensitive, inconsistent, and harsh childrearing, which in turn tended to result in insecure attachment by the child, along with an opportunistic interpersonal orientation. These relationship patterns were associated with aggressive non-compliance in boys and anxious depression in girls. The results of this developmental picture were both somatic and behavioral, represented by earlier maturation and puberty at the somatic level and by earlier sexual activity, less stable heterosexual bonding, and more limited subsequent parental investment at the behavioral level. On the other hand, a family context characterized by harmony and adequate resources had opposite effects at each step in the hypothesized sequence, resulting ultimately in adolescent behavior characterized by later sexual development and more stable, invested bonding with partners and progeny. Belsky, Steinberg, and Draper further argued that the two opposite patterns of adolescent behavior that evolved from the two opposite types of family context each had their own adaptive value with respect to ultimate reproductive success. Specifically, in a harsh, resource scarce world where interpersonal relationships were opportunistic, it made sense to mature rapidly, exit the natal family as soon as possible, and begin the reproductive process with an early opportunity. In contrast, in a secure world characterized by natal family investment of resources, affection, and interpersonal trust, it made sense to delay adolescence and the initiation of reproduction in order to maximize the benefits that could be gained from the natal family.

Although subsequent research on this hypothesized link between experience within the early family environment and adolescent reproductive strategy has had mixed results (Moffitt, Caspi, Belsky, & Silva, 1992; Wierson, Long, & Forehand, 1993; Campbell & Udry, 1995; Graber, Brooks-Gunn, & Warren, 1995), we believe there is sufficient evidence, especially with respect to the central finding that early family environment affects age at menarche, for this hypothesis to be examined further. Furthermore, the hypothesis has two characteristics that serve the broad objective of this book, namely the linking of genetics and reproductive behavior: first, its evolutionary and developmental formulation is highly compatible with a genetic approach; and second, its focus on adolescent sexual behavior and childbearing provides a natural context for examining the acquisition of contraceptive motivation and behavior. However, in order to fully utilize and build upon these two characteristics, we need to integrate two additional perspectives into our theoretical framework, one dealing with a theory of contraceptive behavior and its antecedents and the other dealing with a theory about social bonding and its neural substrate.

2.2   Theory of Contraceptive Behavior

It is useful to distinguish two different aspects of contraceptive psychology: the varieties of contraceptive behavior and the motivational structure that underlies the behavior. The varieties of contraceptive behaviors – that is to say, those behaviors where the intent is to prevent conception – have increased progressively since humans first recognized the connection between sexual intercourse and conception. Abstinence and, probably, withdrawal have been used since the beginning of recorded history. Barrier methods, best represented by the condom, have been used for centuries. In the 1930s, when it was discovered that ovulation occurred between, and not during, menstrual periods, effective periodic abstinence became realizable. The 1960s saw the launching of a new era of chemical contraception which permitted a degree of freedom in sexual behavior that, in turn, stimulated improvements and refinements in all methods, barrier, chemical, and surgical alike.

The complexity of contemporary contraceptive behavior attests to the growing importance of technology in its practice. Despite this complexity, there are two aspects of contraceptive behavior that are central to the use of most methods and that represent hazards for the occurrence of unintended pregnancies, namely the selection of a particular method and the effective use of it. As we have discussed elsewhere (Miller & Pasta, 1996), method selection and effective method use represent two levels of contraceptive decision-making. Method selection represents strategic decision-making in that such choices are made relatively infrequently and are concerned with the individual's overall approach to conception prevention. Method use represents tactical decision-making in that it involves recurrent decisions to use the particular method that has been strategically selected. For example, taking a pill each day or applying a condom with each coitus requires repeated intentional acts, each one of which involves making a decision. As these two examples suggest, perhaps the most important aspect of effective method use is regularity of use.

The motivational structure lying behind these two aspects of contraceptive behavior may be divided into various components, including motivations related to pregnancy and childbearing, abortion, sex and the relationship with the sexual partner, and attributes of the contraceptive methods themselves. Here, we focus on the motivational structure related to pregnancy and childbearing. Further, we will only be concerned with how this motivational structure affects the tactical dimension, regularity of method use.

As already indicated, most contraceptive methods require repeated intentional acts. For this reason, the antecedents of contraceptive behavior may be understood with reference to an intentional systems model similar to the one we have used previously to study the antecedents to proceptive behavior – where the intention is to achieve conception (Miller & Pasta, 1995) – and planned childbearing (Miller, 1994a). Such a model consists of a four-step sequence, which we represent with the acronym TDIB: first, there are childbearing motivations, which are Traits or enduring dispositions to be motivated for and against having children (Miller, 1995); second, these traits are activated in the form of various Desires, such as childbearing, child-number, and child-timing desires; third, these desires are transformed by personal commitment and reality constraints into Intentions; finally, these intentions are implemented through Behavior.

In this paper, as a component within our overall theoretical framework, we will use a trait-desires-intentions model to predict the behavior, regularity of contraceptive use. However, several features of this intentional systems model need clarification and elaboration before it can be adapted from its use in the analysis of proceptive behavior and adequately applied to contraceptive behavior. First, we will conceptualize contraceptive intentions not in terms of the intention to use a method but rather in terms of the intention to prevent pregnancy. We believe this approach captures better the childbearing motivational structure that lies behind the repeated acts which characterize tactical contraceptive decision-making. Second and closely related, we conceptualize contraceptive intentions as being essentially negative in character. That is to say, they are directed at avoiding something rather than achieving something. In previous work (Miller, 1995), we have shown that childbearing motivation has two primary, largely independent dimensions, one positive and the other negative. Proception and contraception reflect that same separation of valency in that they are, respectively, instrumental behaviors where the intention is to achieve (Miller, 1986; Miller and Pasta, 1995) and prevent (Miller and Pasta, 1996) conception. Third and finally, we recognize that any prevention behavior that necessitates repeated intentional acts is highly susceptible to failure at least some of the time. Asked in retrospect to report regularity of use, respondents may not recall specific failures. Nevertheless, they are likely to remember their average level of effort during the period in question. For this reason, we will draw from Bagozzi's (1992) theory of trying and frame our questions to respondents in terms of how much or how hard they have <u>tried</u> to avoid pregnancy.

2.3　Theory of Social Bonding

Up to this point we have discussed material that represents three different periods in an individual's life: experience within the early family environment, puberty and adolescent sexual development, and subsequent (young adult) contraceptive psychology. We are interested in how experience during each of those periods feeds forward to influence that of the same or a subsequent period, with an ultimate interest in its effects on contraceptive behavior. Because we are especially concerned with possible genetic influences, we need a theory that is capable of suggesting biologically-based mechanisms for any observed, feed-forward effects on contraceptive behavior. We believe that a recently articulated theoretical framework regarding the biological basis of childbearing motivation (Miller et al., 1999a) serves that purpose.

In the cited paper, we argue that childbearing motivation is built up through experience from a neural substrate that predisposes humans to be responsive to babies and children. Although originally evolved to generate parental bonding behavior in response to infant stimuli, once the causal relationship between sexual activity and pregnancy was recognized by early hominids, this neural substrate also came to provide the biological underpinnings to behaviors that anticipated that causal relationship, such as proception and contraception. We draw together a body of research evidence that indicates that there are discrete neural centers (nuclei) and tracks located in the ventro-medial aspects of the midbrain, the hypothalamus, and forebrain that participate in the coordination and regulation of maternal behaviors. These neural systems are influenced by gonadal steroid hormones circulating in the blood. Also contributing to their function are a number of specific neurotransmitter systems, including those of dopamine, oxytocin, prolactin, and certain opioid peptides. This complex of neural circuitry, working in conjunction with neural systems dedicated to arousal, perception, and memory, provides the emotional and motivational response that drives nurturant behavior.

We further argue that the neural substrate to parental bonding is part of a more general complex of neural circuitry which is dedicated to the major forms of social bonding that occur during the reproductive life course. We hypothesize four types of neural-based bonding systems, namely succorant, affiliative, sexual, and nurturant systems, all of which share some common circuitry and each of which has unique circuitry. The succorant system bonds the infant and child to the parents. The affiliative system bonds the child, adolescent, and adult to the sibling and peer. The sexual system bonds the adolescent and adult to the opposite sex partner. And the nurturant system bonds the adolescent and adult to the child. Together these four systems work hand in glove to promote the types of bonding necessary for development, survival, and reproduction. Each of these bonding systems has an affectional and a security subsystem, which utilize distinct affects – feelings of warmth or affection and feelings of insecurity or fear, respectively – to accomplish their bonding function.

We believe that these four bonding systems – and the four domains of behavior that they influence – map reasonably well and in potentially instructive ways onto the three life periods that we will be examining in this paper. First, experience with the early family environment reflects both the nurturant input of the parents and the succorant response of the child. Each of these sets of behaviors – nurture and succor – affects and is affected by the other. Siblings and parents also provide affiliative stimulation. Second, puberty and adolescent sexual development, together with its associated romantic experience with opposite-sex partners, reflect the period-typical ascendency of the sexual bonding system, working in close conjunction with the affiliative system. Third, the need to deal with pregnancy risk as well as interest in children and childbearing, during both adolescent development and the young adult period, confronts and tests the strength of the nurturant system. At the same time, romantic partners provide affiliative and sexual stimulation.

Our general premise, then, is that these four bonding systems provide a heuristic framework for understanding developmental connections across the three discussed life periods. An important feature of this framework is the assumption that these bonding systems have evolved to respond in a coordinated way across the life course for adaptive purposes. Such coordination may result from one or more of the following factors: the common effects of these systems' shared neural substrate, the effects that learning has on this shared substrate, or the organizing effects on the common substrate of gene activity modified as a result of some triggering event or set of experiences. In the discussion and interpretation of our results, we will consider each of these factors further.

2.4 The Framework Operationalized

In Figure 1 we schematize the way we have integrated the ideas discussed above in preparation for data analysis. The large boxes represent categories of variable and the arrows connecting those boxes represent predictive relationships of particular interest and relevance to our developmental framework. We are most interested in relationships among boxes 1, 2, and 3, which represent experiences and behaviors within the three life periods we have discussed. As the figure indicates, we expect factors from the early family environment to predict adolescent reproductive development and we expect factors from both of those categories to predict subsequent contraceptive behavior. The factors that we have listed for the early family environment include those that are most frequently cited in the research literature as affecting the timing of adolescent sexual behavior, including family stress, parental affection and support, sexual and other types of abuse, father or mother absence, father's occupation, and mother's reproductive history (Hayes, 1987; Newcomer & Udry, 1984, 1987; Hogan & Kitagawa, 1985; Miller & Bingham, 1989; Stern, Northman, & Van Slyck, 1984; Belsky, Steinberg, & Draper, 1991; Moffitt, Caspi, Belsky, & Silva, 1992; Stevens-Simon & Reichert, 1994; Boyer & Fine, 1992). They also include several factors related to sibship structure (Rodgers, 1983). The factors that we have listed for adolescent reproductive development include a hypothesized sequence leading from age at puberty to lifetime number of pregnancies (Chilman, 1979; Hayes, 1987; Kirby, 1997). The factors that we have listed for contraceptive behavior include, as discussed above, a hypothesized sequence leading from wanting not to get pregnant to regularity of contraceptive use.

Box 4 includes psychological traits. They may be divided into two groups: those that are included because of our theories and those that serve primarily as controls. Aggression and somatic anxiety test Belsky, Steinberg, and Draper's (1991) idea that scarce resources and stress in the early family environment tend to produce externalization (aggression) in boys and internalization (somatic anxiety, our best measure of this construct) in girls. Positive and negative childbearing motivation (PCM and NCM,

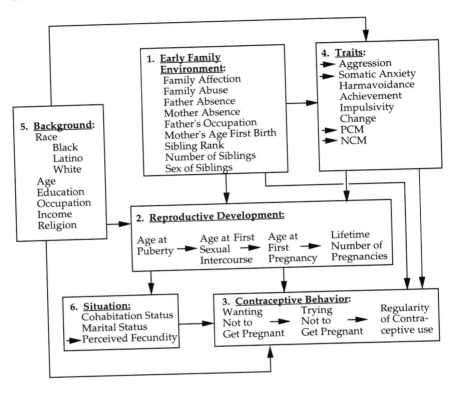

Figure 1. Specific variables and selected hypothesized relationships, arranged according to the theoretical framework. Outcome variables of interest (i.e., those that are endogenous to our model) include all the variables in boxes 2 and 3 and those variables in boxes 4 and 6 that are preceded by arrow heads.

respectively) represent motivational traits and test the model of contraceptive behavior discussed above, which consists of a trait-desires-intentions-behavior sequence. The arrows to each of these four traits indicate that we are interested in their antecedents as well as their consequences. The other four traits represent personality traits that may be expected to affect regularity of contraceptive use and are included as controls in an effort to reduce spurious associations between the factors in boxes, 1, 2, and 3 and regularity of contraceptive use. Here, we are not interested in examining their antecedents.

The factors in boxes 5 and 6 are also included primarily for control purposes and, with one exception, we are again not interested here in modeling their antecedents. The exception is perceived fecundity, the extent to which individuals see themselves together with their partners as capable of achieving pregnancy and birth. Although this is most properly conceptualized as a cognitive factor, we include it here as a situational factor for convenience and because it represents an important element of the couple context of individuals' contraceptive efforts. In addition to its role as an antecedent of contraceptive behavior, we are interested in whether it is affected by adolescent reproductive development.

In the next three sections of this chapter, we describe our methodology for testing the theoretical framework as operationalized above, present the results, and discuss the findings. Because the results suggest an alternative, albeit closely related and complimentary, formulation of our theoretical framework, one that involves distinguishing between implicit and explicit motivations, we turn next to a restatement of the framework. This is followed by three additional sections devoted to the data analytic method used in testing this reformulation, the new results, and a final discussion.

3. METHODOLOGY I

In order to test the theoretical framework, we used data collected from a convenience sample of couples who had requested a pregnancy test in a Planned Parenthood Clinic at one of eight sites located on the San Francisco Peninsula. To be eligible for participation in the study, for which each couple was paid $25, respondents had to be at least 18 years old and involved in a committed relationship for at least the previous six months. One hundred and seventy-eight couples were recruited into the study, with about equal numbers of white, black, and Latino. Of the 178 couples, 76 (43 percent) had a positive pregnancy test. Forty-six carried the pregnancy to term and for them, data collection generally occurred some time in early or mid pregnancy. For the 30 couples who sought an induced abortion, data collection almost always occurred after that procedure. It is important to keep in mind the self-selected nature of our sample and the differential pregnancy and abortion experience of certain subgroups within it because, as we discuss below, these factors dictate caution in accepting the implications of our findings.

Typically, we collected data from both members of a couple at the same time, either in our centrally located research offices, in one of the partner's homes, or at some convenient other location. Each partner was interviewed separately and privately for 45 minutes to one hour. Each partner also completed a group of paper-and-pencil self-report questionnaires either immediately before or just after his or her interview, depending on which partner was interviewed first. For those Latino respondents not completely proficient in speaking and/or reading English, a Spanish-speaking interviewer and/or Spanish language questionnaires were available. Three male partners provided insufficient data, leaving us with a sample size for this study, where no effort is made to model couple interactions, of 178 females and 175 males.

We consider now the measurement of each of the variables shown in Figure 1, beginning with box 1 and ending with box 6. An index of family affection was based on six questions from the family history part of the interview. How loving was your mother? How loving was your father? How much did your mother enjoy being a mother? How much did your father enjoy being a father? How important was religion in your family life? and, How happy was your life together as a family? Response categories for each question varied on a four-point scale from "very" to "not." The index was the mean score on all six questions.

A family abuse index was based on whether any of the following took place in the respondent's family: physical abuse, sexual abuse, emotional abuse, mental illness, and drug or alcohol abuse. For all five types of abuse, the possible responses were yes or no. The index was scored as the total number of yes responses.

Father absence was the sum of whether the father was completely absent from the household during elementary school (no = 0, yes = 1) and high school (same coding). Mother absence was the sum of how much time the mother worked outside the home (none = 1, part-time = 2, full-time = 3) before the respondent was in school, during elementary school, and during high school. Father's occupation was coded according to the U.S. Bureau of the Census (1982) status ratings. Mother's age at first birth, sibling rank, number and sex of siblings were all based on single questions asked during the family history part of the interview.

All four reproductive development variables were based on single questions asked during the sexual, pregnancy, and childbearing history part of the interview. Wanting and trying not to get pregnant were each based on two questions asked during the contraceptive history part of the interview. Their exact form will be reported in the results section. Regularity of contraceptive use was based on an interviewer rating in response to a contraceptive history question about "how regularly or irregularly" the respondent used his or her most recent method. For coitus dependent methods like condoms or foam, the rating scale went from 1. usually (>50% of the time) failed to use method, to 5. never failed to use method. For oral contraception, the rating scale went from 1. missed three or more pills per month, more than two times per year, to 5. never missed a pill. For depo Provera, the rating scale went from 1. six or more months between shots to 5. three months between shots.

Five of the traits were based on self and partner rating of "personality" traits, using a ten-point scale. Each personality trait was described in terms of a cluster of adjectives, derived largely from Jackson's (1984) description of high scorers on his Personality Research Form (PRF) scales. The respondents were then asked to rate both themselves and their partners. For example, the trait of aggression was characterized as follows: is aggressive and assertive; enjoys competition and conflict; may be irritable and argumentative. A subset of respondents also completed parts of the PRF for validation purposes and their scores on the selected scales were used to indicate what was the best weightings of self-rating and rating by partner in constructing the final trait variables. We selected a two-to-one weighting by self and partner as optimal.

Positive and negative childbearing motivation (PCM and NCM) were based on Miller's (1995) Childbearing Questionnaire. Somatic anxiety was based on a physical symptoms check-list, which was modified from the Psychomatic Symptom Checklist (Attanasio, Andrasik, Blanchard, & Arena, 1984). This instrument is based on the finding that patients who check multiple symptoms across multiple organ systems are expressing high levels of anxiety about somatic function.

The background variables were based on general questions at the beginning of the interview that covered such matters as racial-ethnic background, age in years, years of education, current or most recent occupation (again rated according to the Bureau of the Census, 1982), current annual income, and current religion.

Cohabitation was based on a question about whether the respondent and partner lived together and was coded 1 = no, 2 = part time, and 3 = full time. Marital status was determined by a single question and coded 1 = no, not married and 2 = yes, married. Finally, perceived fecundity was based on two questions about how fertile the respondents saw themselves and their partners to be. After defining fertility as the body's ability to get pregnant (for males, make a woman pregnant) and bear a child, the respondent was then asked, "In comparison to other women/men the same age, would you say you are: 1. more fertile, 2. possibly more fertile, 3. the same, 4. possibly less fertile, 5. less fertile." The same question was then asked with reference to the partner and both scores were reverse coded and summed.

In preparation for data analysis, we examined the entire data set and found it relatively free of missing values. Our overall strategy was not to replace missing values for those variables in the three life periods of central interest (boxes 1, 2, and 3 in Figure 1). However, two variables – mother's age at first birth (33 missing) and father's occupation (82 missing) – had relatively large numbers missing. The age variable missing we replaced with the mean for the sample. The occupation variable missing we replaced with an estimated value calculated from four predictors, after first conducting regression analysis in the non-missing part of the sample with respondent education, age, and two race/ethnicity variables as predictors. Of the remaining variables in boxes 1, 2, and 3, only the abuse index and regularity of contraceptive use had a few missing values (see result section). This problem was handled by conducting the entire analysis with pairwise deletions for missing data. For the variables in boxes 4, 5, and 6, the few missing data values were replaced with estimates based on partner data or by sex-specific sample means.

We conducted a linear structural equation analysis of the data, using LISREL (Jöreskog & Sörbom, 1996). With Figure 1 as a guide, we treated all the variables in boxes 2 and 3, plus the five variables in boxes 4 and 6 marked with antecedent arrows, as etas. The remaining variables were treated as ksis and were all allowed to freely correlate. Only two latent variables were used in the model, both with two indicators: wanting and trying not to get pregnant.

We hypothesized a number of predictive relationships between variables in the model on the basis of our theoretical framework and allowed these connections at the beginning of model testing. We outline here our main predictions. We allowed the family affection and abuse indexes to predict age at menarche. We also allowed them, as well as father and mother absence, mother's age at first birth, and number of siblings, to predict age at first sexual intercourse. We allowed the family affection index and mother absence to predict both aggression and somatic anxiety and allowed the latter two traits to predict age at first sexual intercourse. We also allowed the family affection index and mother absence to predict PCM and NCM, both of which in turn we allowed to predict wanting not to get pregnant. We allowed the predictions between the reproductive development and contraceptive behavior variables to be as shown in Figure 1. We allowed age at first sexual intercourse, age at first pregnancy, and lifetime number of pregnancies to predict both wanting and trying not to get pregnant. Finally, we allowed lifetime number of pregnancies to predict perceived fecundity, which we, in turn, allowed to predict trying not to get pregnant.

We conducted the LISREL analysis with separate male and female covariance matrices, which allowed us to test for sex differences. We began with the assumption of no differences (the null hypothesis) and therefore initially constrained all predictors – except those involving age at menarche – to be equal across the sexes, relaxing those constraints only when modification indexes indicated doing so would significantly improve model fit. We relaxed the constraints on any hypothesized predictors that were not significant and then dropped from the model any predictors that were non-significant. If modification indexes indicated that allowing a connection (i.e., adding an unhypothesized prediction to the model) would improve model fit, we did so, first constraining the connection to be equal across the sexes and then, if that was not significant, examining the connection separately in each sex. In general, we only retained connections that were significant with a two-tailed p < .05, although in a few instances we retained connections with a one-tailed p < .05 if they were readily interpretable with the sign observed.

4. RESULTS I

Table 1 gives the means and standard deviations of selected study variables, by sex. The data indicate that the sample is composed of relatively low-income young adults who average only slightly more than a high school education. Not shown in the table is the mixed race/ethnicity of the sample (27% white, 34% black, and 35% Latino), the mixed religious orientation of the sample (33% Protestant, 33% Roman Catholic, and 25% none), and the sample's occupational status rating, which averages around that of a service occupation (4.10) but is higher among females (4.47) than among males (3.72). The couples tend to be living together, at least part time, but only about a quarter of them are married.

The family affection and abuse indexes indicate that the respondents recall a moderate to high level of affection in their family of origin and, on average, at least one type of abuse, with females reporting somewhat more than males. On average, their father was absent from the home during either elementary or high school for about one half of the respondents and their mother was absent part-time for work purposes before, during, and after high school.

Age at first intercourse occurred, on average, shortly after the respondents' sixteenth birthday but was over a year later for the women compared with the men. Age at first pregnancy occurred five and one half years later for the men but just over three years later for the women. For the

Table 1. Mean values and standard deviations of selected variables by sex.

Variables (Scale Range)	Males (N=175)		Females (N=178)	
	Mean	SD	Mean	SD
Age	27.23	6.25	25.55	5.55
Annual Income	19.70	15.19	11.20	12.03
Education	12.53	2.06	12.84	2.03
Family Affection Index (6-24)	19.39	3.81	18.79	4.14
Family Abuse Index[a] (0-5)	1.09	1.28	1.36	1.38
Father Absence (0-2)	0.47	0.92	0.49	0.93
Mother Absence (3-9)	6.36	2.14	6.66	2.04
Father's Occupation (1-6)	3.84	1.34	3.79	1.37
Mother's Age at First Birth	20.88	3.93	20.74	3.73
Age at Menarche	--	--	12.87	1.67
Age at First Sexual Intercourse	15.79	3.17	16.88	2.62
Age at First Pregnancy	21.25	5.45	20.05	4.66
Lifetime Number of Pregnancies	2.26	1.99	2.31	1.95
Cohabitation Status (1-3)	2.50	0.83	2.49	0.84
Marital Status (1-2)	1.27	0.45	1.27	0.45
Perceived Fecundity (2-10)	6.70	1.50	6.74	1.61
Trying Not to Get Pregnant (1-5)	3.27	1.12	3.35	1.02
Rate Trying Not (1-10)	6.39	2.80	6.57	2.74
Wanting Not to Get Pregnant (1-5)	3.72	1.24	3.75	1.23
Rate Wanting Not (1-10)	7.54	2.66	8.03	2.39
Regularity of Contraceptive Use[b] (1-5)	2.67	1.69	2.67	1.76

[a]Male N=163, Female N=167
[b]Male N=169, Female N=177

17 percent of respondents who reported no previous pregnancy, we used their current age for this variable. There is a censoring effect of this coding, which we attempted to minimize by including age as a predictor in the structural equation modelling. Both sexes average somewhat more than two lifetime pregnancies and both perceive themselves as somewhat more fertile than their same age peers. Average parity is just under one child (0.95). Both sexes score higher on wanting not to get pregnant compared with trying not to get pregnant and both score well below the 3.0 mid-point on regularity of contraceptive use.

In Tables 2 and 3 we show the questions, response categories, and response frequencies for the two indicators of both the trying not to get pregnant and the wanting not to get pregnant latent variables. In both cases the first question is categorical and the second scaled, i.e., continuous. (A picture of a 10-point horizontal scale, marked from left to right with low, medium, and high was presented to the respondent as a visual aid.) Table 2 shows a reasonably balanced distribution across the response categories for both indicators of trying not to get pregnant. The distributions in Table 3 show higher frequencies at the high end of wanting not to get pregnant. We cross tabulated the two questions shown in Table 3 and found that although there was a strong positive association between them, response categories 2 and 3 in the first question had almost identical average scores on the second question. For this reason and because these two response categories may be located conceptually at about the same point on the motivational continuum lying between response categories 1 and 5, we combined categories 2 and 3 and used a four point scale from 1 to 4 for purposes of data analysis. In the LISREL analysis measurement model, the respective weightings (i.e., estimated parameters) for the categorical and scaled indicators were 0.84 and 2.21 for the trying not to get pregnant latent variable and 0.74 and 2.28 for the wanting not to get pregnant latent variable.

The overall fit of the final LISREL model was excellent. The chi-square ratio was 0.91, the root mean square error of approximation was 0.0.

We present the final LISREL model in two ways. Table 4 presents the entire structural equation model with unstandardized parameter estimates, t-values, and R^2 values for all prediction pathways, by sex. Variables shown in Figure 1 but not present in Table 4 were not significant predictors for either sex and were dropped from the model, which was then re-estimated. Because our interest focuses on the relationships between the early family environment, adolescent reproductive development, and adult contraceptive behavior, we will not discuss the details of Table 4 but will, instead, present the main findings relevant to these relationships in Figures 2, 3, and 4. In order to facilitate comparison of effects within and across predictor variables, we utilize common metric standardized parameter estimates in these figures.

Table 2. Response frequencies for two interview questions used to construct the Trying Not to Get Pregnant variable. Female version of question is shown.

Which of the following statements was truest of you during the last three months?

1.	I was trying to get pregnant.	6	(1.7%)
2.	I was neither trying to get pregnant, nor trying not to get pregnant.	82	(23.2%)
3.	I was trying not to get pregnant, but not very hard.	125	(35.4%)
4.	I was trying hard not to get pregnant.	77	(21.8%)
5.	I was trying as hard as I could not to get pregnant.	63	(17.8%)
	Total	353	(99.9%)

If you had to rate how much you tried to prevent pregnancy during the last three months, with 1 being not trying at all and 10 being trying as much as possible, how would you rate yourself? [Show scale.]

1 or 2	36	(10.3%)
3 or 4	48	(13.7%)
5 or 6	76	(21.6%)
7 or 8	82	(23.3%)
9 or 10	99	(31.1%)
Total	351	(100.0%)

Table 3. Response frequencies for two interview questions used to construct the Wanting Not to Get Pregnant variable. Female version of question is shown.

The last two questions were about your efforts to avoid pregnancy. The next two questions are more concerned with your feelings about getting pregnant. Which of the following statements was truest of you during the last three months.

1.	I wanted to get pregnant.	11	(3.1%)
2.	I partly wanted to get pregnant and partly wanted not to get pregnant.	77	(21.8%)
3.	I didn't care one way or the other whether I got pregnant.	30	(8.5%)
4.	I did not especially want to get pregnant.	110	(31.2%)
5.	I wanted very much not to get pregnant.	125	(35.4%)
	Total	353	(100.0%)

If you had to rate how much you wanted to avoid pregnancy during the last three months, with 1 being not wanting to avoid pregnancy and 10 being wanting as much as possible to avoid pregnancy, how would you rate yourself? [Show scale.]

1 or 2	15	(4.2%)
3 or 4	24	(6.9%)
5 or 6	62	(17.7%)
7 or 8	65	(19.5%)
9 or 10	185	(52.7%)
Total	351	(100.0%)

Table 4. Structural equation model, showing unstandardized parameter
estimates (t-values) for all prediction pathways, by sex.

Outcome Variable (Male R^2, Female R^2)	Parameter Estimate (t-Value)			
Predictor Variable	Males		Females	
Regularity of Contraceptive Use (0.24, 0.20)				
Trying Not to Get Pregnant	0.96	(8.21)	0.96	(8.21)
Trying Not to Get Pregnant (0.69, 0.54)				
Wanting Not to Get Pregnant	0.74	(10.66)	0.74	(10.66)
Perceived Fecundity	0.10	(4.08)	0.10	(4.08)
Lifetime Number of Pregnancies	-0.05	(-2.48)	-0.05	(-2.48)
Family Affection Index	0.17	(2.98)	0.17	(2.98)
Number of Brothers	0.08	(2.27)	--	--
Occupation	0.14	(3.18)	--	--
Cohabitation Status	-0.09	(-1.90)	-0.09	(-1.90)
Wanting Not to Get Pregnant (0.19, 0.15)				
Perceived Fecundity	0.05	(1.81)	0.05	(1.81)
Age at First Sexual Intercourse	0.05	(3.56)	0.05	(3.56)
Family Abuse Index	-0.08	(-2.59)	-0.08	(-2.59)
NCM	0.01	(2.26)	0.01	(2.26)
White	0.40	(2.95)	--	--
Occupation	0.07	(1.79)	0.07	(1.79)
Cohabitation Status	-0.15	(-3.11)	-0.15	(-3.11)
Perceived Fecundity (0.12, 0.11)				
Lifetime Number of Pregnancies	0.16	(4.15)	0.16	(4.15)
Age at Menarche	--	--	-0.15	(-2.25)
Father Absence	-0.26	(-2.23)	--	--
NCM	-0.02	(-2.90)	-0.02	(-2.90)
Change	0.11	(2.65)	0.11	(2.65)

Table 4. Continued

Outcome Variable (Male R^2, Female R^2) Predictor Variable	Parameter Estimate (t-Value)			
	Males		Females	
Lifetime Number of Pregnancies (0.65, 0.72)				
Age at First Pregnancy	-0.30	(-21.86)	-0.30	(-21.86)
Family Abuse Index	--	--	0.18	(3.32)
Mother Absence	--	--	0.09	(2.51)
PCM	-0.01	(-3.12)	-0.01	(-3.12)
Age	0.26	(22.52)	0.26	(22.52)
Occupation	--	--	-0.26	(-3.17)
Protestant	--	--	-0.41	(-2.65)
Cohabitation Status	0.31	(4.42)	0.31	(4.42)
Age at First Pregnancy (0.42, 0.46)				
Age at First Sexual Intercourse	0.55	(7.60)	0.55	(7.60)
Family Affection Index	0.71	(2.28)	0.71	(2.28)
Father's Occupation	0.86	(3.75)	--	--
Aggression	--	--	-0.29	(-2.23)
Age	0.37	(10.86)	0.37	(10.86)
Occupation	--	--	0.98	(3.52)
Protestant	-1.53	(-2.26)	--	--
Age at First Sexual Intercourse (0.16, 0.33)				
Perceived Fecundity	--	--	0.23	(2.25)
Age at Menarche	--	--	0.23	(2.37)
Family Affection Index	0.62	(3.02)	0.62	(3.02)
Father Absence	-0.29	(-2.10)	-0.29	(-2.10)
Mother's Age at First Birth	0.09	(2.61)	0.09	(2.61)
Number of Brothers	--	--	0.34	(2.58)
Aggression	-0.13	(-2.06)	-0.13	(-2.06)
Change	-0.16	(-2.18)	-0.16	(-2.18)
Harmavoidance	0.19	(2.90)	0.19	(2.90)
Black	-1.19	(-2.41)	--	--
Education	0.35	(5.18)	0.35	(5.18)

Table 4. Continued

Outcome Variable (Male R^2, Female R^2)	Parameter Estimate (t-Value)			
Predictor Variable	Males		Females	
Age at Menarche (---,0.11)				
Family Affection Index	--	--	0.41	(2.26)
Somatic Anxiety	--	--	0.03	(1.77)
Black	--	--	0.64	(2.38)
PCM (0.17, 0.15)				
Family Abuse Index	--	--	1.69	(2.47)
Number of Siblings	--	--	-1.15	(-2.69)
White	-6.02	(-3.69)	-6.02	(-3.69)
Occupation	-1.87	(-2.42)	--	--
No Religion	-7.47	(-4.66)	-7.47	(-4.66)
NCM (0.21, 0.19)				
Lifetime Number of Pregnancies	-0.92	(-2.94)	-0.92	(-2.94)
Mother's Age First Child	--	--	-0.70	(-3.04)
Number of Siblings	-0.93	(-2.83)	--	--
Black	8.23	(6.26)	8.23	(6.26)
Income	-0.12	(-2.60)	-0.12	(-2.60)
Cohabitation Status	2.45	(2.31)	--	--
Aggression (0.05, 0.09)				
Mother Absence	0.16	(3.00)	0.16	(3.00)
Father's Occupation	--	--	-0.26	(-2.47)
Harmavoidance	-0.16	(-3.01)	-0.16	(-3.01)
White	-0.58	(-2.24)	-0.58	(-2.24)
Somatic Anxiety (0.09, 0.08)				
Father Absence	-1.31	(-2.23)	--	--
Mother Absence	0.38	(2.09)	0.38	(2.09)
White	-2.45	(-2.91)	-2.45	(-2.91)
Income	0.09	(3.45)	0.09	(3.45)

Note: A t-value ≥ 1.65 = two-tailed $p \leq .10$; ≥ 1.97 = two-tailed $p \leq .05$; ≥ 2.60 = two-tailed $p \leq .01$.

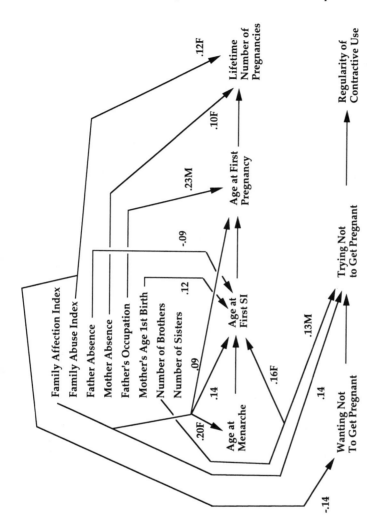

Figure 2. A schematic representation of the structural equation model, showing only the direct effects of early family environment on reproductive development and contraceptive behavior. Numbers represent standardized estimated path coefficients. M and F indicate paths present only in males and in females.

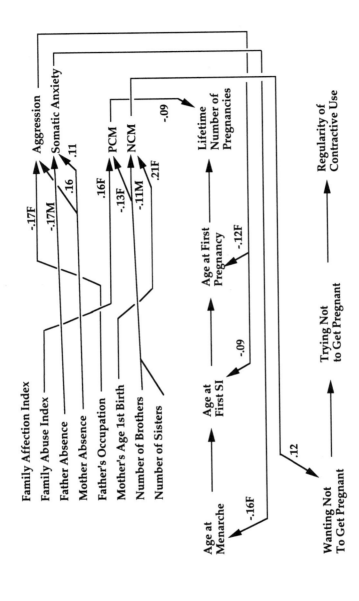

Figure 3. A schematic representation of the structural equation model, showing only the indirect effects through traits of early family environment on reproductive development and contraceptive behavior. Numbers represent standardized estimated path coefficients. M and F indicate paths present only in males and in females.

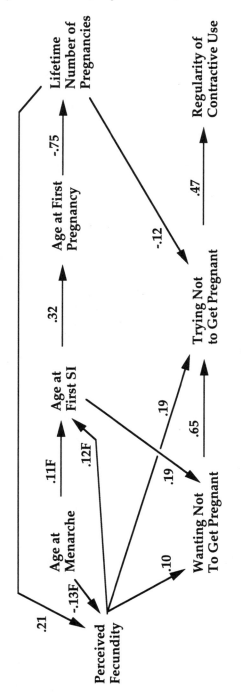

Figure 4. A schematic representation of the structural equation model, showing only the direct effects of reproductive development on perceived fecundity and contraceptive behavior. Numbers represent standardized estimated path coefficients. M and F indicate paths present only in males and in females.

Figure 2 shows the direct effects of the early family environment on reproductive development and contraceptive behavior. The family affection index reveals two types of related effects. First, greater family affection predicts later age at menarche, first intercourse, and first pregnancy, suggesting that parental warmth affects the timing of each step in the adolescent developmental sequence.   Second, greater family affection predicts greater trying not to get pregnant. This effect on adult contraceptive behavior is separate from the developmental sequence effects, suggesting that both types of effects may result from some common antecedent, either biological or psychological. The family abuse index also reveals two related effects.   Greater exposure to abuse in the family of origin predicts more lifetime pregnancies for females and lower wanting not to get pregnant. Consistent with the previously cited literature, both of these effects are pronatal and, again, both may be related to a common biological or psychological antecedent.

Consistent with effects resulting from reduced parental supervision and authority, father absence predicts earlier first sexual intercourse and mother absence predicts more lifetime pregnancies in females.   Consistent with effects that would be expected from children's modeling of parental values and behavior, a higher status father's occupation predicts later age at first pregnancy in males and an older age at mother's first birth predicts older age at first sexual intercourse in both sexes.   Finally, suggesting some sibling learning effects, more brothers predicts an older age at first sexual intercourse for females and trying harder not to get pregnant for males.

Figure 3 shows the indirect effects of the early family environment on reproductive development and contraceptive behavior through intermediate traits.   The indirect effects through aggression and somatic anxiety test Belsky, Steinberg, and Draper's (1991) notion that externalization in boys and internalization in girls is a mechanism through which the stress of early family environment affects adolescent reproductive behavior.   Neither the family affection index nor the family abuse index, the two best candidates for testing that hypothesis, predict aggression or somatic anxiety.   Mother absence predicts both of these traits for both sexes and father absence predicts low somatic anxiety in males.   Also, a higher occupational status of father predicts lower aggression in females.   Although all of these connections are interpretable, they do not lend much support to the tested hypothesis.   Further, although aggression and somatic anxiety do predict earlier adolescent reproductive development in several respects, at least in the case of aggression, the hypothesized specificity for males is not observed.

The indirect effects through positive and negative childbearing motivation shown in Figure 3 test the findings from an earlier study (Miller, 1992) on a relatively affluent, middle-class, married sample of the effects of childhood and adolescent experiences on the development of childbearing motivation. In the earlier study, a variable equivalent to the family affection index was predictive of high PCM and low NCM, and mother absence was oppositely predictive. In this study, neither of those findings is obtained. There was no equivalent to the family abuse index in the earlier study, but here, consistent with the pronatal effects that it demonstrated in Figure 2, the family abuse index predicts higher PCM. Interestingly, number of siblings has a differential sex effect: more siblings decrease PCM in females and NCM in males. Finally, an older mother's age at first birth predicts greater NCM in females, suggesting that mothers who delay their first birth may transmit some values associated with that process to their daughters.

As expected, high NCM predicts greater wanting not to get pregnant but PCM is not predictive, indicating that it is the undesirable features of childbearing, which tend to drive the motivational underpinnings of regularity of contraceptive use. This is very much in concert with previous findings (Miller, 1994b) that high NCM predicts a more accepting attitude toward abortion. The only prediction of PCM is lifetime number of pregnancies but the sign is negative. In this relatively young, largely unmarried sample that has averaged more than two previous pregnancies, this relationship probably reflects the fact that those respondents with high PCM scores and a previous pregnancy have chosen to bear a child, giving them less time to conceive again and making them more vigilant about additional pregnancies.

Figure 4 shows the direct effects of adolescent reproductive development on adult contraceptive behavior and includes perceived fecundity as an intermediate variable. The hypothesized reproductive development and contraceptive behavior sequences behave exactly as expected. Also as expected, more lifetime pregnancies and earlier age at menarche predict greater perceived fecundity, which in turn predicts trying more not to get pregnant. The effect of perceived fecundity on wanting not to get pregnant seems to reflect a motivational effect, which can probably be best understood by considering the low ends of both variables. In other words, low perceived fecundity weakens wanting not to get pregnant because of the wish to prove that perception wrong. Although we need to emphasize again that the retrospective nature of the study may well have a confounding effect, the prediction of an older age at first sexual intercourse by high perceived fecundity suggests that beliefs about one's ability to conceive may moderate, or perhaps even motivate, adolescent sexual risk-taking in females, much as it affects wanting not and trying not to get pregnant in adults of both sexes.

There are two direct connections between the reproductive development sequence and the contraceptive behavior sequence. First, fewer lifetime pregnancies predicts greater trying not to get pregnant. The direction of this prediction is the opposite of what we would expect because it makes sense that those who try more not to get pregnant will have fewer pregnancies. The best interpretation is probably similar to the one discussed above for the relationship between high PCM and fewer pregnancies. In other words, actually bearing a child gives less time for an additional pregnancy and increases contraceptive vigilance, resulting in fewer lifetime pregnancies. Second, an older age at first sexual intercourse predicts a greater wanting not to get pregnant at the time of the study. This connection may reflect a common antecedent, either biological or psychosocial, or it may reflect the effects that delaying first sexual intercourse – and all that is associated with that course of action – has on subsequent reproductive psychology. We will return to a consideration of these possibilities in the next section.

5.   DISCUSSION I

Before drawing any conclusions from our results, it is important to acknowledge three major shortcomings of our data. First, we are studying a convenience sample. It is not possible to determine to what degree self-selection may have influenced the pattern of variable relationships we obtained.   Second, we are examining developmental (longitudinal) relationships using point-in-time (cross-sectional) data. It is quite possible that distortions of memory have influenced the obtained variable relationships. It is also possible that causes and effects are opposite to our interpretations. For example, early menarche may be a cause, not a result of psychological distress (Ge, Conger, & Elder, 1996).

Third, the fact that a subgroup of our sample was pregnant and that this subgroup was further divided into those who did and did not terminate the pregnancy means that responses to some of the questions may have been influenced by changed self-perceptions as a result of these reproductive events. We conducted subgroup analyses (results not shown), comparing mean scores on certain key variables and, indeed, found significant differences. For example, the pregnant subgroup scored somewhat lower than the non-pregnant subgroup on wanting not to get pregnant. However, this difference – as well as the others we examined – probably reflects pre-existing subgroup differences at least as much as it reflects changed self-perceptions. Whatever the impact of these reproductive events may have been, their occurrence, together with the convenience nature of our sample and the retrospective nature of the study itself, indicate that our findings can only be viewed as suggestive and our conclusions as tentative.

We must also acknowledge that our discussion of these tentative findings puts considerable emphasis on the biological part of our theoretical framework.   We recognize that there are many available psychosocial interpretations for the relationships obtained in the LISREL model. Indeed, in the previous section we have made brief reference to a number of them and we will discuss additional ones below. However, both the age at menarche effects that we observed and our bonding framework encourage us to pursue this biological emphasis, which we believe to be very instructive.

We are further encouraged to pursue a biological emphasis by a related consideration. One of the advantages of our multivariate model is that it allows an examination of single variable effects while partialing out the effects of other, related variables. This means that in addition to controlling for possible confounding effects of background factors, as done and found to be important by Mullen, Martin, Anderson, Romans, and Herbison (1996) in their study of the effects of physical, emotional, and sexual abuse in children, we can sort out the interpretation of related variables. For example, with the family affection index in the model, the interpretation of the father and mother absence variables can be more clearly related to loss of parental monitoring and guidance than to loss of affection. By partialing out various psychosocial effects, our multivariate model should increase our chances of identifying effects that have a biological component.

In the discussion that follows, considerable attention will be given to the family affection and family abuse indexes and their interpretation. In preparation for this focus, two methodological points must be emphasized. First, the family abuse index is a composite of three types of interpersonal abuse, mental illness, and drug/ alcohol abuse. There is some evidence that such an additive scale is valid (Hibbard, Ingersoll, & Orr, 1990) but the reader should keep in mind that our abuse construct is very broad, much more so than the sexual and/or physical abuse variables commonly referred to in the literature, not only because we tap five separate types of abuse–related experiences but also because our questions do not specify that the respondents themselves be victims, only that these experiences occurred within the family. Second, by having both the family affection and family abuse indexes present in the model, we are partialing out their separate effects. This is important because it allows us to derive distinct interpretations of these two highly correlated variables, something that is done only infrequently in the research literature (for exceptions, see Weissmann & Silvern, 1994 and Luster & Small, 1997).

We turn now to a closer examination of our core findings and to a more complete interpretation of them in the light of our theoretical framework. This core set of relationships is summarized schematically in Figure 5, where the extensive direct and indirect effects of family affection and abuse on reproductive capacity, motivations and behavior can be seen readily. Family affection has three separate, direct timing effects that involve pubertal changes in the body, sexual/affiliative behavior, and pregnancy (and potentially childbearing). In other words, the amount of affection present in the family seems to affect the full gamut of adolescent reproductive development, suggesting a single mechanism that, because of the effect on pubertal changes, may very well be biological. The direct effect that family affection has on trying not to get pregnant may involve a different, learning-based mechanism. For example, greater love in the home may increase self-efficacy or self-confidence, which then in turn increase efforts to prevent conception. Similarly, the indirect effect of family affection through the connection between age at first sexual intercourse and wanting not to get pregnant may also involve a learning-based mechanism. For example, it can reasonably be argued that delaying sexuality gives the adolescent time to broaden interests, which then compete with the appeal of childbearing and, thereby, strengthen the desire not to get pregnant. However, partly for the sake of parsimony and partly because of our biological focus, we are more inclined to treat both of these effects on adult contraceptive behavior as products of the same mechanism that results in the timing effects. Perhaps this mechanism – about which we will speculate in more detail below – affects some unmeasured feelings about childbearing, which then influence both the desire not to get pregnant and the commitment (and therefore the effort) that is put into trying not to conceive.

Like affection, family abuse has reproductive consequences, but unlike affection, it has no timing effects. This raises the question of whether the effects of abuse involve a different mechanism or whether they simply represent effects of a more extreme form of behavior acting through the same mechanism. The family affection and abuse variables are highly correlated with each other (about – .55 in both sexes) and both have a significant bivariate correlation with most of the variables included in Figure 5, with the pattern of relative strength corresponding well to the differential connections shown in the figure (data not shown). However, we would argue on both theoretical and empirical grounds that family affection (or more correctly for this comparison, the absence of it) and family abuse are not simply two points on a continuum. Theoretically, the former variable measures the amount of recollected love and support in the family, whereas the latter variable measures exposure to various types of exploitation, emotional uncertainty, and danger. In other words, the affection variable is affectionally-based and the abuse variable is fear-based. This conclusion fits well with the affectional and security subsystems of bonding described in the theoretical framework sections. Because we hypothesize and show elsewhere (Miller, Pasta, MacMurray, Muhleman, & Comings, 1999) that these two subsystems have some distinct neural components, this finding in turn suggests that some parts of the neural substrate relevant to the affection and abuse variables are different.

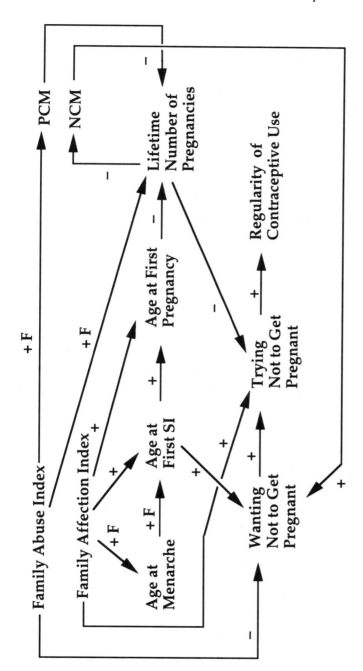

Figure 5. A schematic summary of the core findings.

Empirically, the multivariate results observed in our model indicate that the amount of affection in the family affects the full gamut of reproductive factors, whereas the amount of abuse affects only those factors related to childbearing. Abuse's effects on number of lifetime pregnancies in females and wanting not to get pregnant suggest a common pronatal motivational mechanism. Supporting this interpretation is its direct effect increasing PCM and its indirect effect decreasing NCM in females. (Although PCM seemingly has an anti-natal effect on number of pregnancies, as we suggested in the Results section, this is probably because PCM promotes carrying a pregnancy to term, thus reducing the chance for more pregnancies.) Finally, we note that during the analyses for this paper (results not shown), we identified another pronatal pathway through which abuse may affect childbearing. We observed a trend (one-tailed $p=.098$) for the family abuse index to predict a subscale of PCM (called Joys of Pregnancy, Birth, and Infancy – see Miller, 1995). This subscale in turn strongly predicts lower wanting not to get pregnant.

How can we explain these differences between love and abuse, and more importantly, how can we understand our core findings in terms of processes and mechanisms? To help frame our answer, recall that our theory of social bonding stipulates four major bonding systems that serve to promote pair and social bonding over the reproductive life course. We postulate that this complex of systems (what might be called the bonding suprasystem) has three important features. First, it rests on a neural substrate, with some circuitry being common to all four systems and some being unique to each. Second, it has been designed through evolution to respond to typical bonding requirements as they arise over the life course, with the ultimate goal always being to promote survival and reproduction. These "typical bonding requirements" are addressed by the four components of the life course suprasystem, namely succorance, affiliation, sexuality, and nurturance. Third, the suprasystem has also been designed to respond to typical environmental contingencies. Perhaps the clearest example of this is the capacity an organism has to suppress or release reproductive development and function depending on the situation. For example, among South American marmosets and tamarins, the presence of breeding females tends to suppress the reproductive capability of their adult daughters (Plant, 1994). Similarly, we suggest that exposure to different degrees of affection and abuse within the nuclear family is an environmental contingency to which the bonding suprasystem is prepared to respond.

Figure 6 is a schematic representation of the bonding suprasystem as it operates across the life course and, of necessity, transgenerationally. Let us consider each of its components in terms of our theory of bonding and the core findings reported in the previous section. At the top we show that the nurturant behavior of a parent affects the succorant behavior of a child, which in turn feeds back upon the parent's nurturance in what constitutes an interactive cycle of bonding. There are both affectional and security components of this interaction, affecting both the parent and child. We are suggesting that the family affection index reflects our respondents' retrospective perception of the affectional component and the family abuse index reflects their retrospective perception of the security component. We recognize, of course, that the parent-child interaction is complex and that both parent and child contribute to what becomes the child's perception of the affectional and security attributes of the home environment.

As shown in Figure 6, we postulate that a genetic mechanism responds to the child's perception of affection and security and modulates the development of more adult forms of bonding. Setting aside for the moment a consideration of how that mechanism might actually work, we see in the figure that these perceptions of the family environment affect development of the child's capacity for interactive cycles of bonding that are crucial for successful reproduction. First, there is the sexuality/affiliation cycle, which is central to the establishment of a mating relationship. Second, there is the nurturance/succorance cycle, which is essential for the successful parenting relationship. Although beyond the scope of this paper, it should again be noted that both of these cycles have important affectional and security features.

We have called the genetic mechanism represented in Figure 6 a switch but by that term we do not mean to imply that it is a simple off/on mechanism. Rather we view it as being graded in a way that increases or decreases in a graduated fashion genetic expression, protein formation, and, ultimately, more molar biological response. How might the mechanism work? There are three main possibilities, including mechanisms based on intensity, extensivity, and timing. In other words, the switch might alter the intensity of a response (i.e., how strong it is), its extensivity (i.e., how many systems, behaviors, etc. it affects), or its timing (when it occurs during development). Although at this stage in our understanding of the developmental phenomena being considered here any one of these mechanisms is possible, in our view the simplest and most parsimonious hypothesis at present involves a timing mechanism.

When Belsky, Steinberg, and Draper (1991) first outlined their evolutionary theory of socialization, their emphasis was on the importance of stress in the rearing environment and the effect that it had on puberty and attachment. This emphasis would seem appropriate to the family abuse index findings and suggests that the mechanism of action involves the hypothalamic-pituitary-adrenal (HPA) axis. Some research has indicated that sexual abuse produces HPA axis dysregulation and affects other neuroendocrine stress-response systems (Gerra et al., 1993; Putnam &

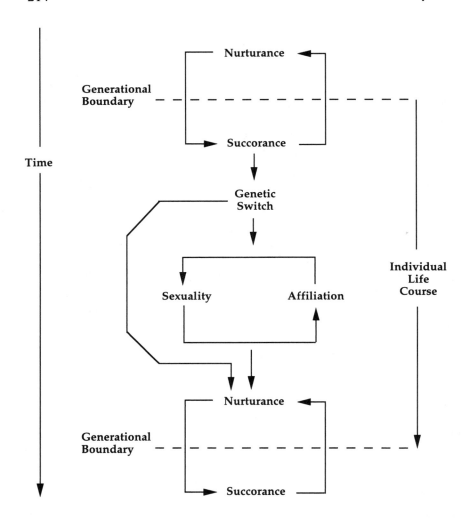

Figure 6. A schematic representation of the hypothesized mechanism underlying the core findings.

Trickett, 1997). However, it is not clear how such dysregulation and other, related effects of stress would lead to an acceleration of puberty, although they could certainly influence several separate developmental pathways (e.g., Wind & Silvern, 1994). Perhaps if the effects of human parental nurturance are thought of as akin to the effects of maternal licking and grooming of rat pups, then some of the neuroendocrine and neurochemical effects that result from variations in rat maternal behavior (Sapolsky, 1997) might play a role. These effects include thyroid hormone release, activation of hippocampal serotonergic projections, increase in glucocorticoid and benzodiazepine receptors, and maintenance of growth hormones.

In our view it is likely that the hypothalamic-pituitary-gonadal (HPG) axis is also involved in the early family environment effects that we have reported, especially those related to the family affection index. The mechanism we propose would operate roughly as follows. Perception of the amount of parental nurturance present in the home would affect a genetic switch that accelerated or decelerated the development of the hypothalamic pulse generator (Plant, 1994). Acting through the HPG axis, and specifically through luteinizing hormone, these pulse generator effects would translate into alterations in the levels of circulating gonadal steroids, including testosterone, estrogen, and progesterone. These steroids, in turn, would act on various neural tissues in the central nervous system (CNS) to affect the expression of both sexual and nurturant (parental) behaviors. The net effect of this proposed mechanism would be, then, to advance or retard the onset of puberty, including both its physical and its motivational and behavioral components.

How parental nurturance in the home might affect a genetic switch, and which specific neurotransmitters might be involved, are open questions. One neurotransmitter that could be central to this mechanism is dopamine. This catechol amine has been shown to play a role in a number of aspects of animal sexual behavior, including humans, especially in males (Melis & Argiolas, 1995). Of particular relevance to this study, polymorphisms of several dopamine receptor genes have been found to be associated with age at first sexual intercourse in humans (Miller, et al., 1999b). This effect may occur through activity in one in particular of the four primary systems of dopaminergic neurons present in the CNS, namely the tuberoinfundibular system which originates in the hypothalamus and projects to the pituitary stalk (Kandel, Schwartz, & Jessell, 1992). The effect may come about through dopamine's role in stimulating the release of gonadotrophic releasing hormone (Plant & Lee, 1995), which controls the release of luteinizing hormone and, thereby, circulating levels of the major gonadal steroids. Further supporting the possible role of dopamine in the proposed genetic switch mechanism is evidence that polymorphisms of this neurotransmitter's receptor genes appear to be associated with positive and negative childbearing motivations (Miller et al., 1999a).

Other neurotransmitters that may be central to this mechanism include oxytocin and the opioid peptides. Evidence suggests that oxytocin plays a role in sexual and parental behaviors and may be central to the process of social bonding (Insel, 1992). Recent studies of the role of dopamine in the formation of partner preference (Yu, Insel, Cescio, & Wang, 1998) indicate that oxytocin and dopamine interact in these bonding processes. In our own work, we have found that polymorphisms of the central oxytocin receptor predict childbearing motivation (Miller, Pasta, MacMurray, Muhleman, & Comings, 1999). Other lines of evidence suggest that one or more of the endogenous opioid peptides participate in the regulation of social bonding (Panksepp, 1997). In the same work just cited, we have demonstrated that polymorphisms of the cannabinoid receptor, which is part of another reward system similar to those of the opioid systems, also predict childbearing motivation (Miller, Pasta, MacMurray, Muhleman, & Comings, 1999). It seems that a number of these bonding related neurotransmitters might participate in a cascade of neurochemical events, jointly constituting what we have called a genetic switch mechanism.

Let us consider, now, how well the process depicted in Figure 6 fits the data of the family affection and abuse indexes. Because the family affection index affects age at menarche, age at first sexual intercourse, and age at first pregnancy independently, a single genetic timing switch affecting pubertal development (not shown in the figure), sexuality, and nurturance would provide the simplest fit to the data for that index. However, because the family abuse index seems to affect only nurturance-related variables in the multivariate context of our model, that simple picture would seem to be incomplete. Perhaps there are two switches, a timing switch and an intensity switch, the former influencing the onset and pace of pubertal and sexual development and the latter influencing the strength of the developing child's readiness to nurture. Alternatively, there may only be a timing switch, one that primarily affects sexuality when it is weakly to moderately turned on but that then begins to affect nurturance as it is more strongly turned on.

Whether it involves one or two switches, an important question that arises is why this two-part arrangement occurs in the first place. One way of thinking about it is as follows. In early family environments, the relative absence of affection is a signal to the developing child to be alert to alternative sources of affection. One good way to achieve such a source (especially once the child is within striking distance of puberty) is to mature earlier and use the sexuality/affiliation bonding cycle to augment and eventually supplant the affectional resources present at home. Turning on sexuality, however, also necessitates preparing the child for the largely inevitable consequence of childbearing by also turning on nurturance. This reasoning suggests that circulating gonadal steroids affect the brain's readiness for nurturance as well as for sexuality. The overall strategy of this sequence involves some risk because it sacrifices a longer growth and development phase, with its attendant consolidation of both physical and mental capacities, and because the alternative sources of affection are – like the child – likely to be less than optimally prepared for the world.

When the developing child experiences fear at the hands of his or her caretakers, the signal is different. It is that any caretakers should be viewed as potential sources of fear; under these conditions the impulse would not be to look for replacement but to bypass alternative sources of affection entirely and go directly to reproduction. This may explain why the experience of childhood sexual abuse seems to foster the desire to conceive among adolescent girls (Rainey, Stevens-Simon, & Kaplan, 1995). It is as though the organism that experiences nurturant fear has less confidence in external affectional sources and concentrates on reproducing itself before it is too late. This strategy is a very high risk one, but it is born out of what the organism perceives as a relatively desperate situation.

When the varieties of affectional deficiency are combined with the varieties of nurturant fear, a mix of adaptive strategies may unfold. If affection is high and fear is low, puberty is delayed and family investment in the child is maximized. If affection is high and fear is high, it appears from the results of several studies (Wind & Silvern, 1994; Luster & Small, 1997) that affection can, to some extent, moderate the effects of fear. If affection is low and fear is low, puberty is accelerated but without an extra push for childbearing. And finally, if affection is low and fear is high, the two systems combine to produce a kind of reproductive one-two punch: find a mate and start reproducing as quickly as possible.

If we step back at this point and look at the mechanism that we have been discussing from the perspective of family adaptation, its useful function becomes even more apparent. Keep in mind that both of our indexes are, in an important sense, family level variables. The family affection index is based on two questions about each parent and two about the family as a whole. Similarly, the family abuse index is based on five types of experiences that may or may not have directly involved the child's parents. We interpret the affectional and fear mechanisms that these indexes reflect to be sensitive barometers of intrafamily conditions. This sensitivity could readily help the family survive in the environment of our evolutionary adaptiveness by promoting a reduction in the dependency burden and helping the child in the family select an optimal strategy for successful reproduction.

The intrafamily conditions to which these mechanisms are sensitive may be divided conveniently into three categories, each operating at a different level of family organization, although we recognize that there may be considerable blurring at the boundaries of these categories.   At the individual level, there is partial loss of a parent through physical or mental illness or complete loss through death or desertion.   At the family level, there are two kinds of factors: those having to do with serious interpersonal conflict, primarily between the two parents or principal caretakers; and those related to demand and conflict growing out of the number of children present in the family.   At the tribal (social) level, there are events and processes that affect family function such as intertribal aggression (war) and social and economic disorganization.   The occurrence of one (or more) of any of these factors can result in a depletion of the affectional resources of the family and/or the development of conditions that induce sustained fear in a child. The mechanism(s) we have identified are capable of responding to these occurrences by cuing the child to hasten reproductive maturation, find a mate and exit the family, and launch his or her own childbearing.   This response reduces the affectional burden on the family and enhances the child's reproductive success.

Of course there are other factors that can affect this developmental process.   The relative abundance of food in the environment can affect the deposition of fat in the body of the developing pre-pubertal child and may thereby affect the timing of puberty.   Recent research (Mantzoros, Flier, and Rogel, 1997) supports the hypothesis that circulating levels of leptin, the hormone released by fat cells, trigger the onset of puberty.   Obviously, this mechanism, and possibly others affecting the timing of puberty, would be expected to interact with the type of affectional and fear-based mechanism(s) we have been discussing.

6.   IMPLICIT AND EXPLICIT MOTIVATION

Up to this point our analysis and discussion has focused on the specific relationships between a set of early family environment variables, a sequence of adolescent development variables, and a sequence of adult contraceptive variables.   Here we change our emphasis and use variables from those three life periods, plus the two childbearing motivational traits, to look at more global changes across the time period we have been considering.   The variables we will use are those previously schematized in Figure 5.   What that figure and the more detailed results on which it is based make evident is that there are connections between the adolescent sequence variables and the contraceptive sequence variables that appear to be motivational.   Further, because of the very nature of the variables involved, these motivational effects appear not to be conscious, in contrast to the effects of NCM.   This observation suggests that making a distinction between implicit and explicit childbearing motivation might be instructive.

Winter, John, Stewart, Klohnen, and Duncan (1998) have recently reviewed and summarized research using the constructs of implicit and explicit motivation. McClelland and his colleagues have been the most articulate advocates for making this distinction (see especially McClelland, Koestner, & Weinberger, 1989). The core of the distinction is the following. Implicit motivations are non-conscious because they are measured by techniques, such as the writing of imaginative stories in response to pictures (the Thematic Apperception Test or TAT, see Atkinson, 1982), in which the respondents are not explicitly describing themselves. Instead, the implicit motivations of the respondents are obtained by the researcher's coding of story content. On the other hand, explicit motivations are conscious because they are measured by questionnaires that ask respondents explicitly about their motivations. Because these types of questionnaires elicit self-attributions, explicit motivations are often referred to as self-attributed motivations. McClelland, Koestner, and Weinberger (1989) detail how implicit and explicit forms of the same motivations (e.g. achievement) are usually not significantly correlated with each other, predict different classes of behavior, and are responsive to different classes of environmental incentives. They conclude that implicit motivations are preverbal, while explicit motivations are more language-based and arise out of what is perceived as important to the self. They further conclude that implicit motivations are mediated by more primitive midbrain structures, while explicit motivations are mediated by higher cerebral cortical structures, especially those that process language. In general, they argue that these two motivational systems operate in parallel, with the more automatic implicit system not particularly involved in making plans or setting goals, but with the explicit system tending to guide implicit motivations into specific channels.

With these ideas in mind, we thought that the adolescent sequence variables, which are developmental and largely behavioral, might well be considered indicators of implicit childbearing motivation in contrast to PCM and NCM, which reflect self-attributed likes and dislikes related to children and child-care. Although implicit motivations are typically measured with verbal reports (story-telling) that are assumed to reflect non-conscious processes, there is no reason to believe that self-reported behavior should be any less successful at revealing such processes. In fact, the frequency of specific behaviors has been proposed as an important way of measuring individual dispositions (Buss & Craik, 1983) and we view the use of variables from the adolescent development sequence to measure a motivational disposition as following in that tradition, with the exception that here we use primarily act timing rather than act frequency as variable indicators.

7. METHODS II

We used the same data described in the previous methods section but limited ourselves to the core variables shown in Figure 5.   We again conducted a linear structured equation analysis using LISREL, but in this case made far greater use of that technique's latent variable capacity.   We conceptualized implicit childbearing motivation as a latent variable with the four reproductive development variables as indicators.   We conceptualized explicit childbearing motivation as a latent variable with PCM and NCM as indicators.   A third latent variable we called perceived family nurturance. The family affection and abuse indexes were indicators for this latent variable, which was hypothesized to predict both the implicit and explicit motivation latent variables.   A fourth and final latent variable was called pregnancy avoidance motivation. It had five indicators drawn from the three variables in the contraceptive behavior sequence (recall that wanting not to get pregnant and trying not to get pregnant were themselves both two-indicator latent variables).   We hypothesized that pregnancy avoidance motivation was predicted by both implicit and explicit motivation variables. As in the previous analyses, we constrained the structural equation model connections to be equal across males and females unless relaxing them improved model fit, but allowed the measurement model connections to vary freely across sex.

8. RESULTS II

Although the latent variable model as specified above produced a good fit, there were signs of instability in the explicit childbearing motivation variable.   Further investigation showed that this was due to the near-zero correlation between PCM and NCM.   Therefore, we abandoned measuring explicit motivation as a latent variable and let PCM and NCM each be an observed variable. When this change was implemented, we found that PCM was neither significantly predicted by perceived family nurturance nor was it a significant predictor of pregnancy avoidance motivation.   Accordingly, it was dropped from the model.   Also, we allowed a covariance between the two wanting not to get pregnant indicators of the pregnancy avoidance motivation.   This connection indicates that these two indicators are more highly correlated than is accounted for by both being in the latent variable.

The overall fit of the resulting LISREL model was excellent.   The chi-square ratio was 1.06 and the root mean square error of approximation was <0.01.

We present the final LISREL model in two ways.   Table 5 presents both the measurement and structural equation models with unstandardized parameter estimates, t-values, and R^2 values for all prediction pathways, by sex. Figure 7 shows a schematic representation of both the measurement and structural equation models, utilizing common metric standardized parameter estimates.

As Figure 7 indicates, in the measurement model by far the strongest indicator for implicit childbearing motivation is age at first pregnancy, a finding that fits well with the meaning of the construct. There is little variation in indicator strength by sex. For perceived family nurturance, the family abuse index is the stronger indicator, twice as strong among females. For pregnancy avoidance motivation, the two continuous versions of the wanting and trying not to get pregnant variables are much the strongest indicators. In the structural equation model, high perceived family nurturance strongly predicts low implicit childbearing motivation and modestly predicts high NCM (which indicates low explicit motivation). High implicit childbearing motivation strongly predicts low pregnancy avoidance motivation. High NCM (or low explicit motivation) weakly predicts high pregnancy avoidance motivation but only in males.

Recognizing that self-report of contraceptive behavior probably contains considerable measurement error in this sample, we decided to add another outcome variable to the model shown in Figure 7: whether or not the respondent's pregnancy test had been positive (coded yes = 1, no = 0). We interpret this variable as an indicator of the effectiveness of contraceptive behavior (although – as discussed in the previous section – we recognize that it may also reflect altered self-perception as a result of the pregnancy) and hypothesized that pregnancy avoidance motivation should predict this outcome variable. We also added perceived fecundity as a predictor of this outcome variable. When the model was re-estimated with these changes, both pregnancy avoidance motivation and perceived fecundity predicted a positive pregnancy test with unstandardized estimated coefficients (t-values) of -.17 (-4.23) and .03 (1.78), respectively. No other variables in the structural equation model predicted this outcome variable.

9. DISCUSSION II

The findings indicate that our measure of implicit childbearing motivation is far more important in the determination of pregnancy avoidance motivation than either of our measures of explicit motivation. This is very much in contrast to our earlier studies, where we found that these explicit measures were major predictors of childbearing desires and intentions (Miller, 1994). Although we did not develop and utilize implicit measures in those earlier studies, we believe that this contrast is important and related to two differences between the current study and our earlier ones: the outcome being predicted and the sample.

Table 5. Latent variable model, showing unstandardized parameter estimates (t-values) for both the measurement model and the structural equation model, by sex.

Measurement Model

Latent Variable	Parameter Estimate (t-Value)			
Indicator	Males		Females	
Perceived Family Nurturance				
Family Affection Index	0.78	(3.26)	0.51	(5.96)
Family Abuse Index	-1.00	--	-1.00	--
Implicit Childbearing Motivation				
Age at Menarche	--	--	-0.27	(-3.16)
Age at First Sexual Intercourse	-1.00	--	-1.00	--
Age at First Pregnancy	-2.00	(-5.24)	-1.92	(-6.64)
Lifetime Number of Pregnancies	0.50	(4.54)	0.60	(5.54)
Pregnancy Avoidance Motivation				
Wanting Not to Get Pregnant, Categorical	0.69	(5.75)	0.67	(5.18)
Wanting Not to Get Pregnant, Continuous	2.29	(6.11)	1.88	(5.22)
Trying Not to Get Pregnant, Categorical	1.10	(6.43)	1.05	(5.86)
Trying Not to Get Pregnant, Continuous	2.41	(6.13)	3.23	(5.98)
Regularity of Contraceptive Use	1.00	--	1.00	--

Table 5. Continued.

Structural Equation Model

Outcome Variable (Male R^2, Female R^2)	Parameter Estimate (t-Value)			
Predictor Variable	Males		Females	
Implicit Childbearing Motivation (.16, .37)				
Perceived Family Nurturance	-1.02	(-5.89)	-1.02	(-5.89)
NCM (.02, .04)				
Perceived Family Nurturance	2.38	(2.78)	2.38	(2.78)
Pregnancy Avoidance Motivation (.20, .17)				
Implicit Childbearing Motivation	-0.18	(-4.69)	-0.18	(-4.69)
NCM	0.01	(1.90)	--	--

Note: A t-value ≥ 1.65 = two-tailed $p \leq .10$; ≥ 1.97 = two-tailed $p \leq .05$; ≥ 2.60 = two-tailed $p \leq .01$. An allowed covariance between the first two indicators of Pregnancy Avoidance Motivation in the theta epsilon matrix = 0.78, with a t-value = 5.21. Structural equation model parameter estimates were constrained equal across gender groups unless not doing so improved model fit.

In our earlier studies, we were primarily interested in how our respondents planned pregnancies, including their number and timing. Those respondents were relatively affluent and white-collar, and all were married. In the current study, we have been interested in how our respondents avoid pregnancies. These respondents are generally low-income and blue-collar, and most are unmarried. It seems reasonable that conscious, self-attributed motivations would play a large role in the few birth decisions that each

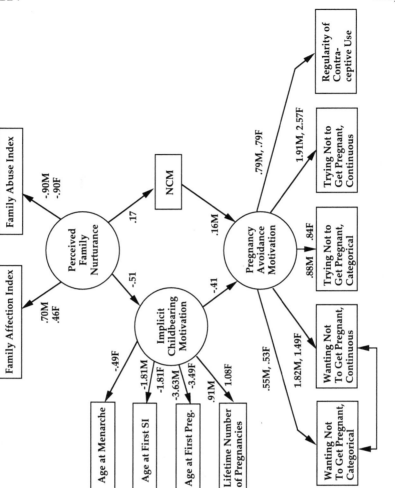

Figure 7. A schematic representation of the measurement and structural equation model, showing the effect of early family environment on reproductive motivations. Numbers represent standardized estimated path coefficients. M and F indicate male and female coefficients, where estimated separately.

couple makes in a lifetime, whereas motivations that are not available to conscious articulation would play a large role in the regulation of repeated intentional acts that are designed to prevent conception. This is very much in line with McClelland's (1989) contention that explicit motivations predict choice behavior, whereas implicit motivations predict "spontaneous behavioral trends over time." It also seems reasonable that explicit motivations would be important among couples where life is relatively well organized and childbearing is normative and that implicit motivations would come to the fore among couples where life is much more chaotic and childbearing is inopportune.

The effect of perceived family nurturance on the two types of motivation is of interest for two reasons. First, the nurturance variable is weighted more heavily by abuse than by love, especially in females. This suggests that fear more than affection determines our respondents' perception of nurturance in their families of origin, perhaps a result of their lower class origins. Apparently this effect is stronger in girls, perhaps because of their greater physical vulnerability. Second, the effect of the nurturance variable on NCM is the opposite of what we found in our earlier study (Miller, 1992). There, although we had no measure of the abuse component, high nurturance produced low negative childbearing motivation. Again, this difference in the effect of nurturance probably reflects a sample difference. If we assume a strong continuity in social class between our two groups of respondents now and when they were living in their childhood families, then our findings suggest that in middle class families nurturance increases the explicit motivation to have children, whereas in lower class families it decreases it. This difference may be related to the greater presence of abuse in lower class families but it may also indicate that in lower class families there is another process at work. Nurturant parents probably promote a variety of interests and activities, some of which compete with and reduce their children's childbearing aspirations. This means that their children develop higher negative childbearing motivation. Higher NCM in turn decreases their children's childbearing desires, on the one hand, and increases their pregnancy avoidance desires, on the other. Ultimately this results in smaller (and later born) families, facilitates the accumulation of resources, and quite probably, increases upward social mobility. If true, this scenario represents an interesting and potentially important example of how a mental mechanism – response to family nurturance – that evolved during the Pleistocene in one environment affects adaptation in another environment, that of a class-stratified society.

Because our measure of implicit motivation is somewhat of a departure from the more typical TAT-related measures, it is worth asking how our type of measure might relate to other types. In our view, there may be a number of different types of implicit childbearing motivations. Our measure is based on the pace of adolescent development and, as we said in the previous discussion, may involve the turning on of a genetic switch that accelerates the development of sexual and reproductive interest patterns. Other types of implicit childbearing motivations may be based on different kinds of experiences (e.g., with childcare or in educational institutions) and be quite independent of the implicit motivation measured here. Recently, Adler and her colleagues have developed a more traditional projective measure of implicit childbearing motivation (Adler & Tschann, 1993; Speier, Melese-d'Hospital, Tschann, Moore, & Adler, 1997) and have shown that it predicts adolescent contraceptive behavior net of the effect of explicit motivations and other demographic and psychological variables. It would be of interest to see whether implicit motivation as measured by our developmental (historical) construct predicts behavior directly or only indirectly through implicit motivation as measured by Adler's more contemporaneous construct.

It should be noted that behaviors as well as motivations can be unconscious. This means that when reporting their contraceptive behavior to us, our respondents are probably not aware of – and therefore do not report – some of what they do. These "implicit" behaviors may be very important in determining the outcomes of interest, namely unintended pregnancies. We included regularity of contraceptive use as an indicator in the pregnancy avoidance motivation latent variable because we believed that a part of it probably does measure such motivation. Some part of it probably also measures explicit behavior, and for this reason it is noteworthy that no variables in the first section analysis predict it directly except trying not to get pregnant. But what about the implicit behavior that it fails to measure? If one accepts the pregnant/not pregnant variable that we added to the model shown in Figure 7 as a proxy for this implicit behavior, then our analysis shows that pregnancy avoidance motivation predicts that otherwise-unmeasured implicit behavior, as it should. Interestingly, neither implicit childbearing motivation nor either of our explicit measures directly predict that behavior. Both funnel their effects indirectly through pregnancy avoidance motivation.

10.   FINAL COMMENT

We asked at the beginning of this paper how genes could be related to individual variation in contemporary contraceptive behaviors. Among the variables in our developmental model, a number are probably subject to genetic influence. There is considerable evidence that variation in age at menarche has a strong genetic component (Meyer, Eaves, Heath, & Martin, 1991; Kaprio et al., 1995). Subfecundity and infertility have many biological roots, some of which may be genetically based (e.g., Legro, Dietz, Comings, Lobo, and Kovacs, 1994). To the extent that perceived fecundity is an accurate assessment of actual fecundity, then it too probably reflects some genetic influence. In work presented elsewhere in this volume (Miller, Pasta, MacMurray, Muhleman, & Comings, 1999) and in previous publications (Miller et al., 1999a), we have found evidence that both positive and negative childbearing motivations are genetically influenced. In addition to these variables, a number of others that affect contraceptive behavior but fall outside of the motivational framework used here – for example, specific personality traits (Miller & Pasta, 1996) – are also good candidates for genetic influence.

Although all of these enumerated variables are of considerable interest, we believe it is the genetic switching mechanism described in this paper that is potentially of greatest importance and that most justifies the paper's title. This mechanism appears to be contingently responsive to conditions of affection and fear within the family. If individuals vary in their sensitivity to such conditions, the switch mechanism may be subject to individual genetic variation as well. Further, this mechanism appears to have a timing effect, and perhaps an intensity effect, within the sexual and nurturant domains of the developing adolescent. Finally, the result of this mechanism is an apparent implicit or nonconscious motivation that moderates efforts to avoid pregnancy by young adults. In short, a switching mechanism that evolved during the Pleistocene to be responsive to family functioning and to enhance reproductive success continues to operate implicitly today in the context of a class-stratified society.

In any future research on the developmental motivational models proposed and explored in this paper, it will be important to avoid the flaws discussed above related to sampling and retrospective report. Ideally, several cohorts of children would be studied longitudinally, beginning prior to puberty and extending at least into late adolescence. We know virtually nothing about how childbearing motivations, desires, and intentions, as well as their pregnancy avoidance counterparts, develop during that period and even less about how that development is affected by intrafamily variables such as available affection, exposure to nurturant fear, and a host of other factors, only some of which we have been able to include in this study. Multivariate model building with longitudinal data from a larger, more representative sample would allow a systematic testing of the ideas put forward here.

ACKNOWLEDGEMENTS

This research was supported by National Institutes of Health Grant R01-HD33251.

REFERENCES

Adler, N.E., & Tschann, J.M. (1993). Conscious and preconscious motivation for pregnancy among female adolescents. In A. Lawson & D. Rhode (Eds.), *The Politics of Pregnancy: Adolescent Sexuality and Public Policy* pp. 144-158). New Haven: Yale University.

Atkinson, J.W. (1982). Motivational determinants of thematic apperception. In A. Stewart (Ed.), *Motivation and Society* (pp. 3-40). San Francisco: Jossey-Bass.

Attanasio, F., Andrasik, F., Blanchard, E.B., & Arena, J.G. (1984). Psychometric properties of the SUNYA revision of the psychosomatic symptom checklist. *Journal of Behavioral Medicine, 7*, 247-258.

Bagozzi, R.P. (1992). The self-regulation of attitudes, intentions, and behavior. *Social Psychology Quarterly, 55*, 178-204.

Belsky, J. (1996). Attachment, mating and parenting: An evolutionary interpretation. *Human Nature, 8*, 61-81.

Belsky, J. (1997, June). *Conditional and alternative reproductive strategies: Individual differences in susceptibility to rearing experience.* Paper presented at the annual meeting of the Human Behavior and Evolution Society, Tucson, AZ.

Belsky, J., Steinberg, L., & Draper, P. (1991). Childhood experiences, interpersonal development, and reproductive strategy: An evolutionary theory of socialization. *Child Development, 62*, 647-670.

Boyer, D., & Fine, D. (1992). Sexual abuse as a factor in adolescent pregnancy and child maltreatment. *Family Planning Perspectives, 24*, 4-11.

Buss, D.M., & Craik, K.H. (1983). The act frequency approach to personality. *Psychological Review, 90*, 105-126.

Campbell, B.C., & Udry, R. (1995). Stress and age at menarche of mothers and daughters. *Journal of Biosocial Science, 27*, 127-134.

Chilman, C. (1979). *Adolescent Sexuality in a Changing American Society: Social and Psychological Perspectives.* Washington, D.C.: USDHEW, PHS, NIH.

Ge, X., Conger, R.D., & Elder, G.H. Jr. (1996). Coming of age too early: Pubertal influences on girls vulnerability to psychological distress. *Child Development, 67*, 3386-3400.

Gerra, G., Caccavari, R., Delsignore, R., Passeri, M., Fertonani Affini, G., Maestri, D., Monica, C., & Brambilla, F. (1993). Parental divorce and neuroendocrine changes in adolescents. *Acta Psychiatrica Scandinavia, 87*, 350-354.

Graber, J.A., Brooks-Gunn, J., & Warren, M.P. (1995). The antecedents of menarcheal age: Heredity, family environment, and stressful life events. *Child Development, 66*, 346-359.

Hayes, C.D. (Ed.) (1987). *Risking the Future: Adolescent Sexuality, Pregnancy, and Childbearing* (Vol. 1). Washington, D.C.: National Academy Press.

Hibbard, R.A., Ingersoll, G.M., & Orr, D.P. (1990). Behavioral risk, emotional risk, and child abuse among adolescents in a nonclinical setting. *Pediatrics, 86*, 896-901.

Hogan, D.P., & Kitagawa, E.M. (1985). The impact of social status, family structure, and neighborhood on the fertility of black adolescents. *American Journal of Sociology, 90*, 825-855.

Insel, T.R. (1992). Oxytocin – a neuropeptide for affiliation: Evidence from behavioral, receptor audiographic, and comparative studies. *Psychoneuroendocrinology, 17*, 3-35.

Jöreskog, K.G., & Sörbom, D. (1996). *LISREL 8: User's Reference Guide*. Chicago, IL: Scientific Software International.

Kandel, E.R., Schwartz, J.H., & Jessel, T.M. *Principles of neuroscience*, Third edition. Norwalk, CT: Appleton and Lange.

Kaprio, J., Rimpela, A., Winter, T., Viken, R.J., Rimpela, M., & Rose, R.J. (1995). Common genetic influences on MBI and age at menarche. *Human Biology, 67*, 739-753.

Kirby, D. (1997). *No Easy Answers: Research Findings on Programs to Reduce Teen Pregnancy*. Washington, D.C.: The National Campaign to Prevent Teen Pregnancy.

Legro, R.S.,. Dietz, D.W., Comings, D.E., Lobo, R.A., & Kovacs, B.W. (1994). Association of dopamine D$_2$ receptor gene haplotypes with anovulation and fecundity in female hispanics. *Human Reproduction, 9*, 1271-1275.

Luster, T., & Small, S.A. (1997). Sexual abuse history and number of sex partners among female adolescents. *Family Planning Perspectives, 29*, 204-211.

Mantzoros, C.S., Flier, J.S., & Rogol, A.D. (1997). A longitudinal assessment of hormonal and physical alterations during normal puberty in boys. V. Rising leptin levels may signal the onset of puberty. *Journal of Clinical Endocrinology and Metabolism, 82*, 1066-1070.

McClelland, D.C., Koestner, R., & Weinberger, J. (1989). How do self-attributed and implicit motives differ? *Psychological Review, 96*, 690-702.

Melis, M.R., & Argiolas, A. (1995). Dopamine and sexual behavior. *Neuroscience and Biobehavioral Reviews, 19*, 19-38.

Meyer, J.M., Eaves, L.J., Heath, A.C., & Martin, N.G. (1991). Estimating genetic influences on the age-at-menarche: A survival analysis approach. *American Journal of Medical Genetics, 39*, 148-154.

Miller, B.C., & Bingham, C.R. (1989). Family configuration in relation to the sexual behavior of female adolescents. *Journal of Marriage and the Family, 51*, 499-506.

Miller, W.B. (1986). Proception: An important fertility behavior. *Demography, 23*, 579-594.

Miller, W.B. (1992). Personality traits and developmental experiences as antecedents of childbearing motivation. *Demography, 29*, 265-285.

Miller, W.B. (1994a). Childbearing motivations, desires, and intentions: A theoretical framework. *Genetic, Social, and General Psychology Monographs, 120*, 223-258.

Miller, W.B. (1994b). The relationship between childbearing motivations and attitude toward abortion among married men and women. *Family Planning Perspectives, 26*, 165-168.

Miller, W.B. (1995). Childbearing motivation and its measurement. *Journal of Biosocial Science, 27*, 473-487.

Miller, W.B., & Pasta, D.J. (1995). Behavioral intentions: Which ones predict fertility behavior in married couples? *Journal of Applied Social Psychology, 25*, 530-555.

Miller, W.B., & Pasta, D.J. (1996). The relative influence of husbands and wives on the choice and use of oral contraception, a diaphragm, and condoms. *Journal of Applied Social Psychology, 26*, 1749-1774.

Miller, W.B., Pasta, D.J., MacMurray, J., Chiu, C., Wu, S., & Comings, D.E. (1999a). Genetic influences on childbearing motivation: A theoretical framework and some empirical evidence. In L.J. Severy & W.B. Miller (Eds.), *Advances in Population: Psychosocial Perspectives* (Vol. 3, pp. 53-102). London, Jessica Kingsley.

Miller, W.B., Pasta, D.J., MacMurray, J., Chiu, C., Wu, S., & Comings, D.E. (1999b). Dopamine receptor genes are associated with age at first sexual intercourse. *Journal of Biosocial Science, 31*, 43-54.

Miller, W.B., Pasta, D.J., MacMurray, J., Muhleman, D., & Comings, D.E. (1999). Genetic influences on childbearing motivation: Further testing a theoretical framework. In J.L. Rodgers, D.C. Rowe, & W.B. Miller (Eds.), *Genetic Influences on Human Fertility and Sexuality*. Mahweh, MA: Kluwer.

Moffitt, T.E., Caspi, A., Belsky, J., & Silva, P.A. (1992). Childhood experience and the onset of menarche: A test of a sociobiological model. *Child Development, 63*, 47-58.

Mullen, P.E., Martin, J.L., Anderson, J.C., Romans, S.E., & Herbison, G.P. (1996). The long-term impact of physical, emotional, and sexual abuse of children: A community study. *Child Abuse and Neglect, 20*, 7-21.

Newcomer, S., & Udry, J.R. (1984). Mothers' influence on the sexual behavior of their teenage children. *Journal of Marriage and the Family, 46*, 477-485.

Newcomer, S., & Udry, J.R. (1987). Parental marital status effects on adolescent sexual behavior. *Journal of Marriage and the Family, 49*, 235-240.

Panksepp, J. (1998). *Affective Neuroscience: The Foundations of Human and Animal Emotions*. New York: Oxford University Press.

Plant, T.M. (1994). Puberty in primates. In E. Knobil & J.D. Neill (Eds.), *The Physiology of Reproduction*, (Vol. 2). New York: Raven Press.

Plant, T.M., & Lee, P.A. (1995). *The Neurobiology of Puberty*. Bristol, UK: Society for Endocrinology.

Putnam, F.W., & Trickett, P.K. (1997). Psychobiological effects of sexual abuse: A longitudinal study. *Annals of New York Academy of Sciences, 821*, 150-159.

Rainey, D.Y., Stevens-Simon, C., & Kaplan, D.W. (1995). Are adolescents who report prior sexual abuse at higher risk for pregnancy? *Child Abuse and Neglect, 19*, 1283-1288.

Rodgers, J.L. (1983). Family configuration and adolescent sexual behavior. *Population and Environment: Behavioral and Social Issues, 6*, 73-83.

Sapolsky, R.M. (1997). The importance of a well-groomed child. *Science, 277*, 1620-1621.

Speier, P., Melese-d'Hospital, I.A., Tschann, J.M., Moore, P.J., & Adler, N.E. (1997). Predicting contraceptive vigilance in adolescent females: A projective method for assessing ego development. *Journal of Adolescent Health, 20*, 14-19.

Stern, M., Northman, J.E., & Van Slyck, R.M. (1984). Father absence and adolescent "problem behaviors": Alcohol consumption, drug use, and sexual activity. *Adolescence, 29*, 301-312.

Stevens-Simon, C., & Reichert, S. (1994). Sexual abuse, adolescent pregnancy, and child abuse: A developmental approach to an intergenerational cycle. *Archives of Pediatric and Adolescent Medicine, 148*, 23-27.

U.S. Bureau of the Census. (1982). *1980 Census of Population: Classified Index of Industries and Occupations* (Final Edition). Washington, D.C.: U.S. Government Printing Office.

Wang, Z., Yu, G., Cascio, C., Liu, Y., Gingrich, B., & Insel, T. R. (1999). Dopamine D2 receptor-mediated regulation of partner preferences in female prairie voles. (Microtus ochrogaster): A mechanism for pair bonding? *Behavioral Neuroscience, 113*, 602-611.

Wierson, M., Long, P.J., & Forehand, R.L. (1993). Toward a new understanding of early menarche: The role of environmental stress in pubertal timing. *Adolescence, 28*, 913-924.

Wind, T.W., & Silvern, L. (1994). Parenting and family stress as mediators of the long-term effects of child abuse. *Child Abuse and Neglect, 18*, 439-453.

Winter, D.G., John, O.P., Stewart, A.J., Klohnen, E.C., & Duncan, L.E. (1998). Traits and motives: Toward an integration of two traditions in personality research. *Psychological Review, 105*, 230-250.

GENETIC INFLUENCES ON HUMAN FERTILITY AND SEXUALITY: COMMENTARY ON CHAPTERS 6-9

Linda Mealey

The four chapters of this section all address a specific instance of a question that psychologists have grappled with ever since Darwin: "To what extent are individual differences attributable to heritable genetic differences, and to what extent are they attributable to species-wide lawful responses to differential experience?" Today debate over this question rages most strongly between behavior geneticists and evolutionary psychologists, but in fact, disagreement over the answer (and over which methods are best to reach an answer), has been a major cause of divisiveness throughout the history of psychology (Cronbach, 1957; Jaynes, 1969; Buss 1984).

The specific instance that serves as the basis for debate in three of these four chapters is the finding that girls who experience various forms of family conflict tend to reach puberty at an earlier age than girls who do not. Belsky and colleagues have approached this topic from a life history perspective, using panel correlations and other longitudinal designs to try to identify temporal relationships among variables; Rowe and colleagues have used family studies to try to partition variance in pubertal age into genetic versus shared environmental causes; Doughty and Rodgers tackle the possibility that non-shared, within-family experiential differences might also contribute to developmental differences. The fourth chapter, by Miller and Pasta, extends this multivariate dynamic into adulthood, using developmental variables as causal inputs into a structural equations model of adult reproductive behavior.

The methodological triangulation demonstrated by this collection of studies is impressive. Unfortunately, when particular genotypes are routinely segregated into particular environments (as is the case when parents of different genotypes create different environments for their children), none of these techniques can clearly distinguish between heritable and facultative (contingent) developmentally-canalized differences (Freedman & Gorman, 1993; Jacobs, 1981; Mealey, 1990). The relationship between family stress and pubertal age can, therefore, still be plausibly explained by either a "genetic" or an "environmental" model; as Rowe notes, no study has yet been able to provide a definitive test.

Before suggesting a new methodological approach for working through this problem, I would like to point out that it is not necessarily the case that observed variation must be explained either by genetic differences or by experiential differences. It is not even the case that variance can always be explained by the sum of these factors. None of the designs discussed in these chapters is able to assess the role of gene-environment interactions, i.e., the differential response of different genotypes to different environments. Because there are no adoption (cross-fostering) studies reported, it is impossible to actually test the model that Belsky postulates, i.e., that some genotypes are more susceptible to environmental influence than others.

Belsky does cite cross-fostering studies of monkeys that support his notion. I might also remind readers of the classic research by Dan Freedman on the differential response to different rearing conditions of dogs of four different breeds (Freedman 1958; see also Segal, Weisfeld & Weisfeld, 1997). In that study, some dogs were relatively indifferent to rearing conditions (e.g., basenjis remained impulsive whether reared under an indulgent or a restrictive regime), while others (e.g., beagles) were more "susceptible" to environmental influence, developing different "personalities" and behaviors under the two rearing conditions.

I suspect that Belsky is correct and that gene-environment interactions are actually quite common. I, too, have argued that heritable differential sensitivity to the environment is likely to be an underlying factor in the maintenance of adaptive variation in humans; specifically, that compared to most of us, psychopaths have a genotype that is relatively unresponsive to early rearing conditions (Mealey, 1995; see also Lykken, 1995). Fuller (1983) noted that this explanation of phenotypic variance is most compatible with the continuous distributions and moderate heritabilities that are exhibited by most personality traits, yet simple additive models are still the norm.

In humans, a true experimental test of the interaction model is clearly impossible. There are, however, several quasi-experimental methods that can be used to distinct advantage.

First, there is a clear role for adoption studies. While neither sampling nor placement is strictly random in adoptions, and despite that adoptive homes include a smaller range of environments than do non-adoptive homes, it is still possible to estimate interaction effects in adoption designs. Modern-day adoptions are different from those of the past, but they are no less rich a scientific resource. Segal (1997) has even compiled a sample of adoptees she refers to as "pseudotwins": same-age, genetically-unrelated children reared in the same household. Studies of adoptive pseudotwins can provide estimates of both within-family shared environment and gene-environment interaction effects.

Second, Crawford and Anderson (1989) have argued that in order to isolate the effect of different environments on development and thus identify adaptive, environmentally-contingent strategies, researchers must control for genetics by studying differences between monozygotic (identical) twins reared apart (MZAs). In the one study of life history variables that has taken this approach, Mealey & Segal (1993) found that in MZA co-twin pairs where one twin had been raised by kin and the other by non-kin, the twin reared by kin was significantly more likely than his or her co-twin to report family conflict and more likely to report parental encouragement of dating. These family dynamics were, in turn, related to age at first marriage and birth of first child (for men) and number of marriages, desire for children, and number of children (for women). Although age at menarche was not part of this analysis, the results did suggest that the reproduction-related behavior of genetically identical individuals differs depending on the family dynamics of the rearing environment.

A third approach is the quasi-experimental monozygotic co-twin control study (Mealey, Bridgstock & Townsend, 1999, Phelps, Davis & Schartz, 1997). MZAs are few and far between and, given the paucity and unusualness of the cases, life history variables cannot be studied prospectively- as is really necessary. On the other hand, monozygotic twins reared together (MZTs) are fairly common and, in line with the findings of Doughty & Rodgers, most research shows that non-shared environmental factors contribute more to the development of individual differences than do shared environmental factors (Plomin & Daniels, 1987, Rowe & Plomin, 1981). This means that the study of differences between MZTs might provide as much useful information as the study of differences between MZAs. Indeed, with MZTs, longitudinal studies can begin at very early ages, providing better, more detailed, and more objective information on the differential experiences of young twins. I believe that co-twin difference studies will be our most useful methodological tool for assessing the role of differential experience on life history strategies (Mealey, in press).

No matter what we determine to be the source(s) of phenotypic variation, a second question implicit in all four of these chapters is whether such variation is in any way adaptive and, if so, does it map cleanly (and adaptively) onto environmental variance. We tend to think of different environments and parenting styles as having statistical main effects that are either good or bad; likewise, we think of particular adolescent and adult outcomes as either good or bad. Instead, we might need to think again in terms of interactive effects: some parenting styles may be "good" in some environments (and/or for some genotypes) but be "bad" in (or for) others.

We also need to ask ourselves "adaptive for whom?" Outcomes based on social criteria (such as the costs to society of teenage pregnancy), do not map onto outcomes using evolutionary criteria. We can expect that natural selection has led to genotypes that produce phenotypes that behave in ways that statistically, over the generations, maximize reproductive success, but we have no reason to expect that during human evolutionary history there was any straightforward connection between parenting style (or attachment, or happiness, or emotional stability) and reproductive success. What we tend to think of nowadays as "poor" and "inconsistent" parenting may have been (and indeed may still be) the best parenting from a purely evolutionary perspective that permits reproductive success as its sole criterion.

The true evolutionary value of life history variation may be impossible to assess in the context of modern society. The processes that produced adaptive phenotypes in the past may not do so today (Crawford, 1998). However, there are, once again, methods that can help provide us with clues. The variance of heritable traits that are evolutionarily adaptive is more likely to be maintained by genetic dominance, epistatic effects, correlated gene complexes, and gene-environment interactions than by purely additive genetic effects (Bailey, 1998, Bruell, 1964,1967). Heritable adaptive traits might, thus, be identified by systematic discrepancies between the "broad heritability" estimates provided by twin studies and the "narrow heritability" estimates provided by adoption studies (Bailey, 1998, Mealey, in press, Segal & MacDonald, 1998): The larger the difference between the two, the greater the evidence for adaptive value.

The questions raised in these four chapters are among the most interesting in all of the social and biological sciences.... and now, at the turn of the twenty-first century, we are finally on the threshold of finding the answers.

REFERENCES

Bailey, J.M. (1998). Can behavior genetics contribute to evolutionary behavioral science? In: C. Crawford and D.L. Krebs (Eds.), *Handbook of evolutionary psychology: Ideas, issues, and applications*. Mahwah, NJ: Lawrence Erlbaum.

Bruell, J.H. (1964). Inheritance of behavioral and physiological characters of mice and the problem of heterosis. *American Zoologist, 4,* 125-138.

Bruell, J.H. (1967). Behavioral heterosis. In: J. Hirsch (Ed.), *Behavior-genetic analysis.* N.Y.: McGraw-Hill.

Buss, D. M. (1984). Evolutionary biology and personality psychology. *American Psychologist, 39,* 1135-1147.

Crawford, C. (1998). Environments and adaptations: Then and now. In: C. Crawford & D.L. Krebs, (Eds.), *Handbook of evolutionary psychology: Issues, ideas, and applications.* Mahwah, NJ: Lawrence Erlbaum.

Crawford, C.B. & Anderson, J.L. (1989). Sociobiology: An environmentalist discipline? *American Psychologist, 44,* 1449-1459.

Cronbach, L.J. (1957). The two disciplines of scientific psychology. *American Psychologist, 12,* 671-684.

Freedman, D.G. (1958). Constitutional and environmental interactions in rearing of four breeds of dogs. *Science 127,* 585-586.

Freedman, D.G. & Gorman, J. (1993). Attachment and the transmission of culture: An evolutionary perspective. Journal of Social and Evolutionary Systems, 16, 297-329.

Fuller, J.L. (1983). Sociobiology and behavior genetics. In: J.L. Fuller & E.C. Simmel, (Eds.), Behavior genetics: Principles and applications. Hillsdale, NJ: Lawrence Erlbaum.

Jacobs, J. (1981). How heritable is innate behaviour? *Zeitschrift fur Tierpsychologie, 55,* 1-18.

Jaynes, J. (1969). The historical origins of 'ethology' and 'comparative psychology'. *Animal Behaviour, 17,* 601-606.

Lykken, D. (1995). *The antisocial personalities.* Mahweh, NJ: Erlbaum.

Mealey, L. (1990). Differential use of reproductive strategies by human groups? *Psychological Science 1,* 385-387.

Mealey, L. (in press). Kinship: The tie that binds (disciplines). In P. Davies & H. Holcomb (Eds), *Evolution and minds: Psychological and philosophical perspectives,* Boston: Kluwer.

Mealey, L., Bridgstock, R. & Townsend, G.C. (1999). Symmetry and perceived facial attractiveness: A monozygotic co-twin comparison. *Journal of Personality and Social Psychology, 76,* 157-165.

Mealey, L. & Segal, N.L. (1993). Heritable and environmental variables affect reproduction-related behaviors, but not ultimate reproductive success. *Personality and Individual Differences, 14,* 783-794.

Phelps, J.A., Davis, J.O. & Schartz, K.M. (1997). Nature, nurture and twin research strategies. *Current Directions in Psychological Science 6,* 117-121.

Plomin, R. & Daniels, D. (1987). Why are children in the same family so different from each other? *Behavioral and Brain Sciences, 10,* 1-16.

Rowe, D.C. & Plomin, R. (1981). The importance of nonshared environmental influences on behavioral development. *Developmental Psychology, 17,* 517-531.

Segal, N.L. (1997). Same-age unrelated siblings: A unique test of within-family environmental influences on IQ similarity. *Journal of Educational Psychology, 89,* 381-390.

Segal, N.L. & MacDonald, K.B. (1998). Behavior genetics and evolutionary psychology: A unified perspective on personality research. *Human Biology, 70,* 157-182.

Segal, N.L., Weisfeld, G.E. & Weisfeld, C.C., Eds. (1997). *Uniting psychology and biology: Integrated perspectives on human development.* Washington, DC: APA Press.

Chapter 10

THE GENETICS OF HUMAN SEXUAL ORIENTATION

Khytam Dawood and J. Michael Bailey

Key words: male homosexuality, sexual orientation, sexual identity, familiality, heritability, genetics

Abstract: The primary focus of this chapter is on the evidence to date regarding the population genetics of sexual orientation. The bulk of the available evidence suggests moderate heritability for male sexual orientation. Female sexual orientation has been studied much less extensively, but recent studies are consistent with a genetic contribution for women as well (Pattatucci & Hamer, 1995). Familial aggregation has been observed in nuclear family studies of both male and female homosexuality, although the genetic and environmental contributions to this familial clustering have not been resolved by the main twin studies published thus far, which have yielded somewhat contradictory results. Recent molecular genetics studies will also be reviewed, including the two main strategies that have been employed so far - linkage and association analysis.

1.   INTRODUCTION

Several excellent reviews have been written in recent years on the biological bases of sexual processes. The bulk of the available evidence suggests moderate heritability for male sexual orientation. Female sexual orientation has been studied much less extensively, but recent studies are consistent with a genetic contribution for women as well (Pattatucci & Hamer, 1995). Familial aggregation has been observed in nuclear family studies of both male and female homosexuality, although the genetic and environmental contributions to this familial clustering have not been resolved by the main twin studies published thus far, which have yielded somewhat contradictory results. Recent studies have shown homosexual males to have an excess of homosexual brothers (Pillard and Weinrich, 1986; Bailey et al., 1991), and homosexual females to have an excess of homosexual sisters (Pillard, 1990; Bailey & Benishay, 1993), compared to heterosexuals of the same sex. A finding of familiality suggests that a trait is influenced by genetic and/or shared environmental factors; however the family study methodology cannot resolve the question of which is operating. In addition, the issue of whether male and female homosexuality are co-familial remains unresolved. The largest family study of sexual orientation to date (Bailey & Bell, 1993) failed to demonstrate independent familiality for male and female homosexuality, and this is noteworthy, since with respect to sample size, it is the most powerful test to date.

In the current review, we will provide an overview of the major studies conducted thus far on the genetics of human sexual orientation, including family and twin studies. We will also discuss the implications of recent advances in molecular genetic studies.

As currently practiced, behavior genetics research proceeds, roughly, in three stages. First, studies determine if a trait or characteristic runs in families. Familial aggregation suggests, but does not prove, a genetic contribution to the characteristic, because some traits (e.g., religious affiliation, last names) run in families for environmental reasons. In order to disentangle genetic and familial environmental effects, more sophisticated designs are necessary, including twin and adoption studies. Finally, given positive family and genetic findings, studies then attempt to elucidate the genetic contribution. These would include, for example, studies of cofamiliality (e.g., do male and female homosexuality run in the same families) and potential indicators of genetic loading (e.g., do feminine gay men have more gay brothers than other gay men). Studies exist from all three stages of research regarding both male and female sexual orientation.

2.   FAMILY STUDIES

The first contemporary family-genetic study of sexual orientation was Pillard and Weinrich's (1986). They recruited homosexual and heterosexual male probands using newspaper advertisements that did not mention the nature of the study. Index subjects (or probands) were interviewed about their own sexuality as well as their siblings' sexual orientations. They obtained permission to contact, and successfully contacted, the large majority of probands' brothers, in order to verify proband reports. Results suggested that homosexual probands were quite accurate at assessing their siblings' sexual orientations (provided that they expressed a high degree of confidence, which they were typically able to do). Most important, gay male probands had an excess of gay brothers (22%) compared to heterosexual male probands' brothers (4%).

Subsequent studies have used similar methodologies, with one exception. The study with a very different methodology, by Bailey et al., (1999) recruited a sample of gay and bisexual men from consecutive admissions at an HIV outpatient center. (In most studies, both gay and bisexual men have been included as "homosexual" probands.) The most important aspect of such an ascertainment strategy is that it is more systematic than advertising for volunteers, and it may be less subject to self-selection biases. All available studies have focused on the rate of homosexuality in siblings rather than other first degree relatives (e.g., parents or offspring), due to the decreased reproduction of homosexual people.

Table 1 contains the results of recent family studies. Considering results for male probands, the rate of homosexuality among brothers has, with one exception (Pillard and Weinrich, 1986) hovered closely around 9%. These rates have exceeded those for heterosexual controls as well as the prevalence estimates from recent large-scale epidemiological surveys, suggesting that male homosexuality is familial. Homosexual women also appear to have more homosexual sisters than do heterosexual controls, though the familiality estimates have varied more widely for women.

Table 1 also contains information concerning the cofamiliality of male and female homosexuality. Do male and female homosexuality run in the same families? There is a trend in Table 1 for gay male probands to have more gay brothers than gay sisters, and for the opposite pattern to obtain for lesbian probands, suggesting that at least some of the familial factors influencing male homosexuality differ from those influencing female homosexuality. However, the largest study to date (Bailey and Bell, 1993) did not find this pattern. The degree of cofamiliality of male and female homosexuality remains a very open question.

Table 1. Rates of Male and Female Homosexuality, Nontwin Siblings in Recent Studies.

Study	Criterion For Sibling Homosexuality	Brothers		Sisters	
		Pro-bands	Con-trols	Pro-bands	Con-trols
Male Probands					
Pillard & Weinrich, 1986	Sibling report, Kinsey score of 2-6	.22	.04	.08	.09
Bailey et al., 1991	Subject's rating with certainty	.10	.00	.02	.00
Bailey & Pillard, 1991	Subject's rating with certainty	.09		.06	
Bailey & Bell, 1993	Subject's rating of sibling, Kinsey score 2-6	.09	.04	.03	.01
Bailey et al., 1999	Subject's rating with certainty	.09		.04	
Female Probands					
Pillard, 1990	Sibling report, Kinsey score of 2-6	.13	.00	.25	.11
Bailey & Benishay, 1993	Subject's rating with certainty	.07	.01	.12	.02
Bailey et al., 1993	Subject's rating with certainty	.05		.14	
Bailey & Bell, 1993	Subject's rating of sibling, Kinsey score 2-6	.12	.00	.06	.01

Table entries are percentages. Missing table entries indicate that studies did not assess the respective rate.

3. GENETIC STUDIES

The most elegant technique for studying the genetics of human behavior is the study of monozygotic (MZ) twins reared apart. Any similarity between such twins must be attributable to their (perfect) genetic similarity. Unfortunately for science, separated twins are rare. There are approximately 200 in the world literature. Furthermore, if one is studying a relatively uncommon phenotype such as homosexuality, the number of separated MZ twins with the phenotype is rare indeed. There have been 6 such pairs reported in the literature (Eckert et al., 1986). Both male pairs were concordant (i.e., similar) for adult homosexual feelings and behavior (though one pair differed in their degree of homosexual inclination). Although it is tempting to dismiss this finding as based on too few subjects, it is a statistically very unlikely result. If we liberally assume the prevalence of male homosexuality to be 4% (Gebhard, 1972), the probability that both pairs would be concordant is less than one in five hundred. In contrast, all 4 female pairs were discordant (i.e., different) for sexual orientation. These results have been taken by some to imply that female sexual orientation is less heritable than male sexual orientation. That conclusion is premature because the sample size is too small to enable a statistically powerful test. (Because the male finding was significant, power was obviously sufficient.) Furthermore, even if the same familial-genetic factors influenced both male and female sexual orientation, one would expect lower female concordances based on the lower base rate of female homosexuality, which appears to be about half as common as male homosexuality (Gebhard, 1972). A definitive test of separated MZ female twins would probably require more than 10 pairs in which at least one twin is lesbian. Due to the rarity of separated MZ twins, such a sample is unlikely ever to be assembled. If human behavior genetics required separated twins, we would need to resign ourselves to ignorance.

Fortunately, other more practical designs exist. The most common methodology used by contemporary human behavior geneticists to disentangle genetic and environmental determinants is the study of unseparated twins. This methodology compares the similarity of monozygotic and dizygotic (DZ) twins who have been reared together. MZ twins are more similar, genetically, than are DZ twins. Because both kinds of twins have been reared together, environmental similarity is assumed to be equal (more about this assumption later). Thus, if MZ twins are more similar than DZ twins, this reflects their greater genetic similarity and is evidence that genetic factors influence the phenotype.

The first twin study of (male) homosexuality, by Kallmann (1952), ascertained gay twins in the "homosexual underworld" and correctional/mental institutions of New York City. Remarkably, 100% of 37 MZ twin pairs were concordant compared to 15% of 26 DZ pairs.

Kallmann's (1952) study had a number of methodological defects, including its over reliance on (evidently) mentally ill gay men, lack of information on zygosity diagnosis, and especially its anomolously high rate of MZ concordance compared to other studies (Rosenthal, 1970). Still, it is remarkable that despite its promising results, nearly forty years passed before another large twin study of male homosexuality was attempted. Several twin studies have been conducted in recent years, and their results are given in Table 2. In our discussion we focus on the four largest studies (Bailey and Pillard, 1991; Bailey et al., 1993; King & McDonald, 1992; Whitam et al., 1993), reported in recent years.

Table 2. Concordance Rates for Twin Studies of Homosexuality.

Study	MZ Concordance	DZ Concordance
Male Studies		
Kallmann (1952)	100 (37/37)	15 (3/26)
Heston & Shields (1968)	60 (3/5)	14 (1/7)
Bailey & Pillard (1991)	52 (29/56)	22 (12/54)
Buhrich et al. (1991)	47 (8/17)	0 (0/3)
Female Studies		
Bailey et al. (1993)	48 (34/71)	16 (6/37)
Combined Male and Female		
King & McDonald (1992)	25 (5/20)	12 (3/25)
Whitam et al. (1993)	66 (25/38)	30 (7/23)

Table entries are percentages (N concordant/total).

3.1   Bailey, Pillard, et al., 1991, 1993

Homosexual probands were recruited via advertisements in gay or lesbian publications (e.g., "Do you have a twin or an adoptive brother?"). We recruited 2 kinds of probands: probands with twins or probands with adoptive brothers or sisters (for the male and female studies, respectively). Adoptive siblings were raised with the probands but are genetically unrelated to them. Probands were interviewed, especially concerning the sexual orientations of their twins. We contacted probands' twins when possible, and confirmed that probands were quite accurate in assessing their twins' sexual orientations.

In our male study, 52% of the MZ cotwins were also gay or bisexual, compared to 22% of the DZ cotwins and 11% of the adoptive brothers. In our female study 48% of the MZ cotwins were also lesbian or bisexual, compared to 16% of the DZ cotwins and 6% of adoptive sisters. Thus, for both men and women, the rates conformed to a partially genetic model, with highest concordance in the most genetically similar (MZ) group, and lowest concordance in the least similar (adoptive) group.

We attempted to assess the magnitude of genetic influence via genetic modelling (Neale & Cardon, 1992). The degree of genetic influence is quantified as heritability, the proportion of phenotypic variance that is attributable to genetic variance. Thus, heritability ranges from 0 (no genetic influence) to 1 (complete genetic influence). In order to calculate heritability from studies such as ours, it is necessary to make several assumptions. First, we assumed that although we measured sexual orientation on a dichotomous scale (i.e., heterosexual versus homosexual), the underlying causal structure was dimensional. That is, genetic influences are polygenic (i.e., numerous genes each with small effect), and environmental influences are similarly multifactorial. This corresponds to a multifactorial threshold model (Reich et al., 1975). Second, it is necessary to assume a population base rate for homosexual orientation. We considered a range of possibilities from 4% to 10% for males, and from 2% to 10% for females. (In light of recent studies, even the lower figures are probably too high. However, using more conservative figures would not have affected heritability estimates very much.) Third, because we suspected that ascertainment methods led to concordance-dependent bias, we assumed different degrees of such bias, from none at all to the case in which probands from concordant pairs were three times more likely to be ascertained than probands from discordant pairs. For both men and women, heritability estimates ranged from approximately .30 to approximately .70. Because of our uncertainty regarding important assumptions, advocating more precise estimates would be inappropriate.

3.2   Other studies

Whitam et al., (1993) studied 38 pairs of MZ twins (34 male pairs and 4 female pairs), obtaining a 66% concordance rate compared to a 30% rate for 23 DZ pairs (14 male pairs, 9 opposite sex pairs, no female pairs). In the other relatively large recent study, King and McDonald (1992) failed to find significantly different concordance rates in 20 mixed-sex MZ and 25 DZ pairs (25% and 12%, respectively). It is noteworthy that King and McDonald's MZ rate was lower than Whitam et al.'s DZ rate. It is possible that Whitam et al.'s sample relied more heavily on word-of-mouth sampling, thus inflating the representation of concordant pairs. (Concordant pairs are more likely to be known in the gay and lesbian communities because they are seen together, etc.)

3.3   Methodological Issues

The most common criticism aimed at the studies listed in Table 2 concerns the "equal environments assumption" that the trait-relevant environment is no more similar for MZ twins than for DZ twins or adoptive siblings. A frequent objection to human twin studies (e.g., Lewontin, Rose, and Kamin, 1984) is that parents treat MZ twins especially similarly and that this similar treatment, rather than the twins' similar genotype, could explain their similar behavior. Indeed, MZ twins are more likely to have been dressed alike and to have shared the same room as children, among other things. The question is whether such treatment makes them more similar, and the evidence suggests that this is not the case, at least for traits studied so far (Kendler et al., 1993; Plomin et al., 1990). For example, MZ twins whose parents make an effort to treat them alike do not behave more similarly than do MZ twins whose parents make an effort to treat them differently. MZ twins whose parents mistakenly believe that they are DZ twins are as similar as they should be based on their true zygosity. It is true (and unfortunate) that the equal environments assumption has not been directly studied in the context of sexual orientation, but it is also true that existing evidence does not contradict the equal environments assumption.

A more serious potential problem concerns ascertainment bias. Ideally, one could recruit probands by interviewing every member of a well-defined population of, say, gay men and asking them if they were twins. Psychiatric genetics has been able to ascertain twins systematically by interviewing consecutive psychiatric admissions, but this strategy was obviously unavailable to those studying homosexuality. The problem with ascertaining twins via advertisements is that self-selection factors are likely to distort results. The most likely way in which this would occur is that gay men whose twins are also gay would be more willing to volunteer than gay men with heterosexual twins (e.g., because the latter might fear conflict from their twins). Kendler and Eaves (1989) have called this kind of bias "concordance-dependent ascertainment bias." This type of bias inflates concordances compared to the population rates, though it does not lead to spurious findings of heritability. Spurious findings could be obtained, however, if concordance-dependent bias were stronger for MZ than for DZ twins. Although there is no evidence that this is so, it cannot presently be excluded. A systematic ascertainment strategy is the most crucial methodological goal for future population genetics studies of sexual orientation.

One final limitation of twin and family studies is worth emphasizing. Even accepting its validity, the evidence reviewed so far is uninformative regarding proximate etiological mechanisms. Genetic evidence does not necessarily support a neuroendocrine explanation, for example. One could envision a host of other genetic pathways to homosexuality. For example, a gene for feminine appearance in boys could elicit differential treatment from parents, leading to a homosexual outcome in adulthood. Or a gene for intolerance of intolerable men might lead a woman to adopt a homosexual identity. We hasten to add that we do not believe that either of these examples is likely, and we do in fact think that the neuroendocrine hypothesis is most likely. But available studies cannot distinguish among them. Molecular strategies that can identify specific genes for sexual orientation will be much more useful in elaborating the developmental pathways from genes to behavior.

4. MOLECULAR GENETICS OF SEXUAL ORIENTATION

Recent developments in molecular biology have generated techniques that could yield important scientific breakthroughs. Specifically, such techniques could allow the discovery of specific genes affecting sexual orientation. When such genes are discovered, their functions can be investigated. This would provide a tremendous advance in our knowledge of sexual orientation development. The two main strategies that have been employed so far are association and linkage analysis.

4.1 Association Studies

Association studies explore the relation between genetic variation at a specific locus and phenotypic variation. There have been claims, for example, that alcoholism is associated with the A1 allele of the D2 dopamine receptor gene (Noble & Paredes, 1993), though these are controversial

(Gelernter et al., 1993). Association studies require that one have a very specific hypothesis, in contrast to linkage studies, which may search the entire genome and examine genetic markers rather than genes.

Macke et al. (1993) examined DNA sequence variation in androgen receptor gene, reasoning that some variants may affect sexual differentiation of the brain. They found no significant differences in the distributions of mutations in homosexual and heterosexual men. Furthermore, using linkage analysis, they showed that sibling pairs concordant for homosexuality were no more likely than chance to share the same androgen receptor allele.

4.2   Linkage Studies

Linkage aims to map genes onto chromosomal regions. The logic is simple. If two relatives both have the same phenotype (e.g., are both homosexual), they probably share any gene that contributes importantly to the phenotype. If so, then they have a high likelihood of sharing the larger chromosomal segment that contains the gene.

Hamer et al. (1993) conducted the first genetic linkage study of (male) sexual orientation. Their report consisted of two major analyses: a pedigree study and a linkage study. First, they examined family pedigrees in a "randomly ascertained" sample of homosexual probands. As reported in Table 1, the probands had a high rate of gay brothers, 13.5%. Furthermore, their pedigrees showed an excess of gay uncles and male first cousins on the maternal side compared to the paternal side, though the difference was not significant. This excess was more pronounced in a subsequent analysis of an additional sample of 38 families with two gay brothers. If there are genes for male homosexuality, then these families should be especially rich with them. The probands' maternal uncles and cousins (through maternal aunts; no gay cousins were sons of maternal uncles) had rates of male homosexuality of 10.3% and 12.9%, respectively, compared to rates of 1.5% and 3.1% for paternal uncles and cousins. This pattern of results is precisely what one would expect if an X-linked gene influenced male sexual orientation. In X-linked inheritance, males with the trait inherited the gene from their mothers, and hence have more maternal than paternal relatives with the trait.

Because of the suggestion of X-linkage, Hamer et al. then searched the X chromosome using linkage analysis. Specifically, they looked at the pairs of gay brothers without evidence of paternal transmission (e.g., they excluded a few cases in which the father may have been gay; the pairs analyzed in the linkage study included all eligible pairs analyzed in both pedigree studies). These examined 22 genetic markers distributed across the X chromosome, in order to see if brothers concordant for homosexuality were also concordant for the markers. For chromosomal region, Xq28, at the tip of the long arm of the X chromosomes, 33 of 40 pairs of gay brothers shared all the markers. This was statistically different from the expected rate (20 of 40) suggesting that a gene influencing male sexual orientation lies within that chromosomal region.

Hamer's study has been met both with great interest and skepticism (Risch et al., 1993). The reasons for the interest are fairly obvious. Linkage studies bring one much closer to finding a gene, thus providing definitive proof that such genes exist and eventually allowing discovery of the genetic mechanism distinguishing homosexual and heterosexual men. The scientific importance of such a breakthrough can hardly be exaggerated.

Some skepticism derived from concerns about Hamer et al.'s study. Risch et al. (1993) raised three main issues. First, they suggested that the pedigree finding, that gay men had an excess of gay maternal relatives, could be due to bias. They speculated that people may know more about their mothers' side of the family (presumably because mothers are more socially oriented, on average). Second, they argued that even if the finding of an increased rate of gay maternal relatives were true, it could be due to fertility patterns. Even if a gene for male homosexuality were autosomal (i.e., not X-linked) gay men are unlikely to have inherited it from their fathers, because men with the gene tend to be gay and gay men tend not to have children. Third, Risch et al. argued that Hamer et al. overestimated an important parameter, λ, that reflects the increased prevalence of a trait in first degree relatives compared to the background, or general population, rate. This parameter, which affects probability estimates, has not yet been precisely estimated in a large and careful study. Risch et al. chose values from available studies to yield the lowest plausible value of λ, which would have rendered the linkage analysis statistically nonsignificant. In our view, the concerns raised by Risch et al. are worth the attention of future research, but are not fatal flaws in Hamer et al.'s study.

Other reasons for skepticism have less to do with Hamer et al.'s study than more general concerns about linkage analysis of genetically complex traits. By genetically complex traits, we mean those whose transmission patterns do not fit classic Mendelian patterns such as autosomal dominant or recessive, or X-linked dominant or recessive. All evidence suggests that male sexual orientation is inherited, if at all, in a complex manner. Linkage analysis has provided important breakthroughs for Mendelian traits (Bishop, 1990), but it has also provided some false leads, especially for genetically complex behavioral traits (Risch & Merikangas, 1993). Indeed, to date not a single molecular finding concerning behavior has been widely accepted as valid by the scientific community, and several highly publicized findings have failed to replicate. This is in part because the number of studies examining any one trait has been relatively small. But it could also reflect the likely possibility that genes underlying behavior variation are typically of small effect and thus difficult to detect. Thus, it is especially important that linkage findings be replicated.

To date three replication attempts have been reported. One research group obtained pedigree information from 182 families with at least two gay brothers (Rice et al., 1999). They failed to find a significant excess of gay maternal uncles or cousins. In a subset of 41 sibling pairs, they also failed to replicate the finding of linkage to Xq28. Unlike Hamer et al., however, they did not exclude brother pairs with strong evidence of paternal transmission, and it is unclear how many of their subjects would have met Hamer et al.'s inclusion criteria. Nevertheless, the failure to replicate either of Hamer et al.'s key findings surely diminishes the probability that they are correct.

In contrast, Hamer (1995) reported a successful replication. In this study, Hamer included data from heterosexual brothers as well as gay brothers, and found that brothers' similarity for sexual orientation was statistically related to the sharing of Xq28 markers. That is, not only did gay brothers tend to share the markers, but gay-straight pairs tended not to share the markers. The magnitude of the genetic effect was smaller in Hamer's second study, however, and the result was barely statistically significant.

Because of the conflicting replication results, the status of the Xq28 linkage finding is unresolved. When studies are small, replications count more than failures to replicate. Nevertheless, larger studies will be needed to determine whether male sexual orientation is influenced by a gene in Xq28.

If the twin studies are correct in their findings of genetic influence on sexual orientation, then eventually molecular techniques such as linkage and association analysis will allow us to discover specific genes affecting sexual orientation. For now, however, the promise of such techniques remains largely unrealized, though they have generated one finding (Hamer et al., 1993; Hamer, 1995) that is highly deserving of further study.

Our most recent study (Bailey et al., 1999) is generally consistent with past studies in suggesting elevated rates of homosexuality among siblings of homosexual men. The rates for brothers reported for our two relevant samples, 7.3% and 9.7%, are similar to the median figure of 9% found in other family studies (Bailey & Pillard, 1995). Insufficient control data were collected from heterosexual probands in this study to provide a direct test of the difference in the two rates. However, three recent large surveys of sexual behavior from Great Britain (Wellings, Field, Johnson, & Wadsworth, 1994), France (Spira & al., 1993), and the United States (Laumann et al., 1994) have provided estimates of adult homosexual behavior. These estimates vary sensibly with the stringency of the respective definitions. For men, the least stringent definition examined in the studies, any homosexual experience ever, yielded an estimate of 4.1% in the French survey (Spira & al., 1993) 6.1% in the British survey (Wellings et al., 1994), and 7.1% in the American survey (Laumann et al., 1994). A much more stringent criterion, same-sex activity during the year preceding the survey, yielded rates of 1.1%, 1.1%, and 2.7% for the three samples, respectively. The criterion closest to that explored herein, self-identification as "gay" or "bisexual," was employed only in the American study (Laumann et al., 1994), and applied to 2.4% of men. The 95% confidence interval for the rate of nonheterosexuality among brothers in the HIV study ranged from 5% to 10%; for the rate of homosexuality among brothers in the Gay Pride sample it ranged from 6% to 14%. In a direct comparison of nonheterosexuality among brothers in our samples with the rate of male nonheterosexuality in the American survey (Laumann et al., 1994), both the HIV sample (χ^2=23.5, df=1, p<.001) and the Gay Pride sample (χ^2=33.6, df=1, p<.001) yielded significantly elevated rates. The two rates were not significantly different from each other (χ^2=1.2, df=1, p>.25). Assuming a rate of 2.4% for the population prevalence of male nonheterosexuality, the respective estimates of λs are 3.0 and 4.0. We believe that these are the best current estimates of λs, and they should be considered in power calculations in future linkage studies.

The present estimates for brothers are lower than that for dizygotic (DZ) twins obtained in some recent genetics studies of homosexuality (Bailey & Pillard, 1991; King & McDonald, 1992; Whitam et al., 1993), which range from 12% (King & McDonald, 1992) to 30.4% (Whitam et al., 1993). There are at least two possible explanations. First, twins share some aspects of environment (e.g., prenatal influences and contemporaneous rearing) that siblings do not, and these are plausibly related to sexual orientation. Second, the recent twin studies have recruited probands using advertisements that clearly revealed the nature of the sudies, which probably magnified concordance-dependent ascertainment bias (Kendler & Eaves, 1989). Consistent with this interpretation, recurrence rates for both fathers and sisters appeared to be inflated in the sibling pairs sample, which was also recruited via advertisements. In order to determine if the rate for DZ twins is greater than that for nontwin brothers, it will be necessary to recruit twins more systematically, as in the present study.

Familial clustering of male and female homosexuality has also been found in some previous studies (Bailey & Bell, 1993; Pattatucci & Hamer, 1995), but not in others (Bailey & Benishay, 1993; Bailey, Pillard, Neale, & Agyei, 1993). The most influential biologic theory of sexual orientation is that male homosexuality results from incomplete masculinization of relevant brain structures during prenatal development; female homosexuality has an analogous, opposite, explanation (Ellis & Ames, 1987; Friedman & Downey, 1993; LeVay, 1996). It is, thus, unclear how the same familial factors would affect both male and female homosexuality.

Our pedigree study found no evidence that male sexual orientation is influenced by an X-linked gene. This contrasts with the work of Hamer and colleagues, who found evidence for X-linkage both in pedigree (Hamer et al.

1993) and linkage (Hamer et al. 1993; Hu et al. 1995) studies. Because this study only reported pedigree data, we focus on the differences between it and the original study by Hamer et al. (1993), which also included pedigree data. Both of our relevant samples (N=350 and N=167) were substantially larger than that of Hamer et al. (N= 76), and our sibling pairs sample was also somewhat larger than theirs (65 versus 38 families, respectively). Thus, each of our samples had greater power to detect familial patterns consistent with X-linkage than the corresponding sample in Hamer et al.'s report. We were, however, unable to do so.

Although our results differed from the pedigree data in Hamer et al.'s initial report, they are not strong evidence against X-linkage. For example, consider the HIV sample, in which yielded a rate of 7.3%. Even if all familial aggregation were attributable to an X-linked gene, this would imply a rate of homosexuality of approximately 3.7% among maternal uncles. But that sample obtained a rate of 4.4%, which is even higher than the unrealistic model would predict. The power to detect differences in maternal versus paternal transmission of the magnitude that realistic models would predict is surely low.

One conceivably relevant difference between our studies concerns ascertainment. Our HIV sample was ascertained entirely through an HIV clinic, and our Gay Pride sample was ascertained exclusively at a Gay Pride rally. Hamer et al. ascertained a portion of their random sample through an HIV clinic, but the rest was recruited via two gay mental health clinics. It is possible that gay men recruited from the latter sites are more likely to carry an X-linked gene for homosexuality compared with those from HIV clinics. Sampling bias would not, however, explain the different results of the two studies' sibling pairs samples, which were recruited similarly. Assuming that our samples are representative of gay men in the general population, our data suggest that X-linked genes account for relatively few cases of male homosexuality.

5. SUMMARY

During recent years, evidence suggesting that familial and genetic factors affect male and female sexual orientation has accumulated. The genetic evidence is substantially stronger for male than for female sexual orientation. Although the studies have provided generally consistent results, this consistency is tarnished by a consistency of methodological problems, especially ascertainment bias. Available evidence is strong enough to support efforts to conduct better-controlled and more definitive studies. Genes could well comprise an important determinant of sexual orientation. Furthermore, environmental determinants apparently frequently differ between MZ twins reared in the same family, and evidently cause their effects early in development. Molecular genetics techniques such as linkage hold great promise, but that promise has not yet been realized for behavioral traits.

REFERENCES

Bailey, J. M. (1995). Sexual orientation revolution [news]. *Nature Genetics, 11*, 353-354.

Bailey, J. M., & Bell, A. P. (1993). Familiality of female and male homosexuality. *Behavior Genetics, 23*, 313-322.

Bailey, J. M., & Benishay, D. S. (1993). Familial aggregation of female sexual orientation. *American Journal of Psychiatry, 150*, 272-277.

Bailey, J. M., & Pillard, R. C. (1991). A genetic study of male sexual orientation. *Archives of General Psychiatry, 48*, 1089-1096.

Bailey, J. M., & Pillard, R. C. (1995). Genetics of human sexual orientation. *Annual Review of Sex Research, 6*, 126-150.

Bailey, J. M., Pillard, R. C., Dawood, K., Miller, M. B., Farrer, L. A., Trivedi, S., & Murphy, R. L. (1999). A family history study of male sexual orientation using three independent samples. *Behavior Genetics, 29*, 79-86.

Bailey, J. M., Pillard, R. C., Neale, M. C., & Agyei, Y. (1993). Heritable factors influence sexual orientation in women. *Archives of General Psychiatry, 50*, 217-223.

Bailey, J. M., Willerman, L., & Parks, C. (1991). A test of the maternal stress theory of human male homosexuality. *Archives of Sexual Behavior, 20*, 277-293.

Bishop, J. (1990). *Genome.* New York: Simon & Schuster.

Buhrich, N., Bailey, J. M., & Martin, N. G. (1991). Sexual orientation, sexual identity, and sex-dimorphic behaviors in male twins. *Behavior Genetics, 21*, 75-96.

Eckert, E. D., Bouchard, T. J., Bohlen, J., & Heston, L. L. (1986). Homosexuality in monozygotic twins reared apart. *British Journal of Psychiatry, 148*, 421-425.

Ellis, L., & Ames, M. A. (1987). Neurohormonal functioning and sexual orientation: A theory of homosexuality-heterosexuality. *Psychological Bulletin, 10*, 233 - 258.

Gebhard, P. (1972). Incidence of overt homosexuality in the United States and Western Europe. In J.M. Livingood (Ed.), *National Institute of Mental Health Task Force on Homosexuality: Final Report.* Washington, DC: U.S. Government Printing Office.

Gelernter, J., Goldman, D., & Risch, N. (1993). The A1 allele at the D2 dopamine receptor gene and alcoholism. A reappraisal. *Journal of the American Medical Association, 269*, 1673-1677.

Heston, L. L. & Shields, J. (1968). Homosexuality in twins: A family study and a registry study. *Archives of General Psychiatry, 18*, 149-160.

Friedman, R. C., & Downey, J. (1993). Neurobiology and sexual orientation: current relationships. *Journal of Neuropsychiatry & Clinical Neurosciences, 5*, 131-153.

Hamer, D. H., Hu, S., Magnuson, V. L., Hu, N., & et al. (1993). A linkage between DNA markers on the X chromosome and male sexual orientation. *Science, 261*, 321-327.

Hu, S., Pattatucci, A. M. L., Patterson, C., Li, L., Fulker, D. W., Cherny, S. S., Kruglyak, L., & Hamer, D. H. (1995). Linkage between sexual orientation and chromosome Xq28 in males but not in females. *Nature genetics, 11*, 248-256.

Kallmann, F. J. (1952). Twin and sibship study of overt male homosexuality. *American Journal of Human Genetics, 4*, 136-146.

Kendler, K. S. & Eaves, L. J. (1989). The estimation of probandwise concordance in twins: the effect of unequal ascertainment. *Acta Geneticae Medicae et Gemellologiae, Twin Research, 38*, 253-270.

Kendler, K. S., Neale, M. C., Kessler, R. C., Heath, A. C., & Eaves, L. J. (1993). A test of the equal-environment assumption in twin studies of psychiatric illness. *Behavior Genetics, 23*, 21-28.

King, M. & McDonald, E. (1992). Homosexuals who are twins: A study of 46 probands. *British Journal of Psychiatry, 160*, 407-409.

Laumann, E. O., Gagnon, J. H., Michael, R. T., & Michaels, S. (1994). The social organization of sexuality: Sexual practices in the United States. (pp. 283-320). Chicago: The University of Chicago Press.

Lewontin, R.C., Rose, S., & Kamin, L.J. (1984). *Not in our genes.* New York: Pantheon Books.

LeVay, S. (1996). *Queer science: The use and abuse of research into homosexuality.* Cambridge, Massachussetts: MIT Press.

Macke, J. P., Bailey, J. M., King, V., Brown, T., Hamer, D., & Nathans, J. (1993). Sequence variation in the androgen receptor gene is not a common determinant of male sexual orientation. *American Journal of Human Genetics, 53*, 844-852.

Neale, M.C. & Cardon, L.R. (1992). *Methodology for genetic studies of twins and families.* Dordrecht, Netherlands: Kluwer.

Noble, E.P. & Parades, A. (1993). Recent developments in alcoholism: Molecular biology and behavior. *Recent Developments in Alcoholism, 11*, 345-362.

Pattatucci, A. M. L., & Hamer, D. H. (1995). Development and familiality of sexual orientation in females. *Behavior Genetics, 25*, 407-420.

Pillard, R. C. (1990). The Kinsey scale: Is it familial? In S. A. S. J. M. R. David P. McWhirter (Ed.), *Homosexuality/heterosexuality: Concepts of sexual orientation. The Kinsey Institute series, Vol. 2* (pp. xxvii, 423). New York, NY, US: Oxford University Press.

Pillard, R. C., & Weinrich, J. D. (1986). Evidence of familial nature of male homosexuality. *Archives of General Psychiatry, 43*, 808-812.

Plomin, R., DeFries, J.C., & McClearn, G.E. (1990). *Behavioral genetics: A primer.* New York: W.H. Freeman.

Reich, T. Cloninger, C.R., & Guze, S.B. (1975). The multifactorial model of disease transmission: I. Description of the model and its use in psychiatry. *British Journal of Psychiatry, 127*, 1-10.

Rice, G., Anderson, C., Risch, N. & Ebers, G. (1999). Male homosexuality: Absence of linkage to microsatellite markers at Xq28. *Science, 284*, 665-667.

Risch, N. (1990). Linkage strategies for genetically complex traits. I. Multilocus models. *American Journal of Human Genetics, 46*, 222-228.

Risch, N. & Merikangas, K.R. (1993). Linkage studies of psychiatric disorders. European *Archives of Psychiatry and Clinical Neuroscience, 243*, 143-149.

Risch, N., Squires-Wheeler, E., & Keats, B. J. B. (1993). Male sexual orientation and genetic evidence. *Science, 262,* 2063-2065.

Rosenthal, D. (1970*). Genetic theory and abnormal behavior.* New York: McGraw-Hill.

Spira, A., Bajos, N., Bejin, A., Beltzer, N., & et al. (1992). AIDS and sexual behaviour in France. *Nature, 360*, 407-409.

Wellings, K., Field, J., Johnson, A., & Wadsworth, J. (1994). *Sexual behavior in Britain: The national survey of sexual attitudes and lifestyles.* New York: Penguin.

Whitam, F. L., Diamond, M., & Martin, J. (1993). Homosexual orientation in twins: A report on 61 pairs and three triplet sets. *Archives of Sexual Behavior, 22,* 187-206.

Chapter 11

ENDOPHENOTYPES FOR MARITAL STATUS IN THE NAS-NRC TWIN REGISTRY

Susan L. Trumbetta and Irving I. Gottesman

Key words: marital status, divorce, never marrying, pair bonds, genetic factors

Abstract: This study explored two possible endophenotypes for marital status: 1) the predisposition to form and maintain lasting pair bonds and 2) the predisposition to have multiple mates over the life span. These endophenotypes were constructed using 1972 and 1985 marital status data from a followed-up subsample of the NAS-NRC WWII Veteran Twin Registry. In the 1972 data, consisting of 2297 MZ and 2443 DZ twin pairs, 42% of the variance in pair bonding could be attributed to additive genetic and 58% to nonshared environmental factors and measurement error. Of the variance in multiple mates, 28% could be attributed to additive genetic and 62% to nonshared environment/error factors. In the 1985 data, consisting of 1359 MZ and 1208 DZ twin pairs, 31% of the variance in pair bonding could be attributed to non-additive genetic and 69% to nonshared environment/error factors. Of the variance in multiple mates, 22% could be attributed to additive genetic and 78% to nonshared environment/error factors. Although parameter estimates were marked by wide confidence intervals, no variance in either endophenotype could be attributed to common family environment.

1.   INTRODUCTION

Although statistical evidence suggests that genetic factors contribute to risk for divorce (Jockin, McGue, & Lykken, 1996; McGue & Lykken, 1992; Trumbetta, Gottesman, Turkheimer, & Page, 1999) and for never marrying (Trumbetta et al., 1999), little research explores the pathways of influence between genes and marital status. Psychological phenotypes, such as personality (Jockin et al., 1996) and psychopathology (Trumbetta et al., 1999), account for some of the genetic and environmental variance in marital status. These phenotypes, however, probably influence marital status indirectly, by affecting behavioral patterns that increase the probability of never marrying or of ever divorcing. These unmeasured behavioral patterns that increase probabilities for particular marital status outcomes may be thought of as *endophenotypes* for marital status.

How might such endophenotypes best be described? Marriage resembles pair-bonding patterns found in other species and implicates behaviors related to the formation and maintenance of stable heterosexual partnerships. These behaviors are presumably of adaptive significance in evolution for procreation and for optimal nurturance of offspring (Dunbar, 1995; Komers, 1996; Larsen, 1991; Piper, Evers, Meyer, Tischler, Kaplan, & Fleischer, 1997; Salo, Shapiro, & Dewsbury, 1993). Evolutionary theory also suggests that a variety of mating strategies may be more or less successful under differing environmental conditions (Morell, 1998). Two such mating strategies are 1) strong parental investment and 2) mate diversification. Strong parental investment usually involves a strong pair bond, and this will be reflected in greater likelihood of maintaining a marriage over the life span. Mate diversification, on the other hand, will be associated with a greater likelihood of multiple marriages.

As genetic criteria are used to identify specific disease phenotypes, especially in the realm of psychiatry (McGuffin et al., 1994), genetic criteria can also help to discern which of these two endophenotypes of marital status may be the stronger indicator of genetic variation in mating strategies. The endophenotype showing higher heritability will, in principle, more readily facilitate identification of more fundamental biological endophenotypes for marital status. This paper uses data from the National Academy of Sciences-National Research Council (NAS-NRC) World War II Veteran Twin Registry to evaluate by genetic criteria pair bonding and mate diversification as potential endophenotypes for marital status variation.

2. METHOD

2.1 Participants

The NAS-NRC registry of World War II and Korean Conflict era veteran twins was compiled from a search of birth certificates for live, male, European-American, multiple births in the United States between the years 1917 and 1927. It ascertained 15,924 pairs of twins in which both had served in the U.S. armed forces (Kendler & Robinette, 1983), from which 20,946 individuals, representing 8,747 complete pairs, responded to the initial 1965 questionnaire (Carmelli, Swan, & Robinette, 1991; Fabsitz, Kalousdian, Carmelli, Robinette, & Christian, 1988). The sample under consideration is a subset of this larger registry, whose recruitment and zygosity typing techniques have been described elsewhere (Jablon, Neel, Gurshowitz, & Atkinson, 1967; Kendler & Robinette, 1983). Several epidemiological questionnaires were sent to registry members over the years, and for this study, responses to the second and seventh questionnaires from 1972 and 1985, respectively, were used. A total of 16,519 individuals responded to either of these two questionnaires, with 14,300 individuals responding to the 1972 questionnaire, and 9,395 to the 1985 questionnaire. This yielded 4,960 complete twin pairs of known zygosity in 1972, of whom 4,740 indicated marital status for both twins; and 2,603 complete twin pairs of known zygosity in 1985, of whom 2,567 indicated marital status for both twins.

Because eligibility for the NAS-NRC registry was predicated on both members of a pair having met World War II military service standards, the sample was selected for physical and mental health. An extensive literature considers the effects of selective ascertainment on this registry, particularly for the exclusion of psychiatric illness (Fabsitz et al., 1988; Hrubec, 1973; Kendler, 1986; Kendler & Robinette, 1983). Additional ascertainment biases have accrued, as successive questionnaires sampled only those twins who survived and remained locatable over the years. We assume that respondents were, on average, in better health and more stable in their residence than non-respondents were.

2.2 Measures

Questionnaire data did not indicate whether remarriage followed divorce or the death of a spouse. Because divorce was more frequently represented than widowhood both in the registry and in its primary population cohort in the U.S. census data, responses denoting remarriage were coded as previously divorced. Widowed respondents were coded as if continuously married, on the assumption that spousal death occurs randomly among the married male population. Admittedly, slight differences have been observed between young adult personalities of continuously married and widowed men (Kelly & Conley, 1987), and mortality data suggest slight differences in widowhood rates across socioeconomic classes (LeClere, Rogers, & Peters, 1997). Nevertheless, univariate biometric models showed no discernible differences between the inclusion and exclusion of widowers among the continuously married in this sample, so they were included as right-censored, continuously married persons.

We coded marital status for 1972 and 1985 in ordinal patterns representing two endophenotypes for marital status, stable heterosexual pair bonding and mate diversification. We ranked pair bonding according to increasing levels of "successful" marital involvement. Thus, never marrying ranked lowest and continuous marriage ranked highest. Remarriage ranked lower than continuous marriage because it implied a previous, unsuccessful pair bond, but higher than divorce because it represented a resumption of pair-bonding following divorce. Table 1 present's ordinal values of pair bonding assigned to each marital status for 1972 and 1985 and their frequencies. Clearly, this scale provides only a coarse approximation, as it is derived from a point prevalence of divorce and was limited by the sparse marital information ascertained by the two questionnaires.

Mate diversification was ranked according to the minimum number of marital status changes required for any current marital status. Never marrying required no marital status change from birth; marriage, one change. "Married once and divorced" required two changes. "Multiple marriage, married now," could indicate any number of marital status changes, but a minimum of three could produce such a response, while "multiple marriage, divorced now" required a minimum of four changes. "Separated" was ranked higher than marriage and lower than "married once and divorced." Widowhood was set equal in value to marriage. Ordinal values for marital instability based on 1972 and 1985 marital status categories and their frequencies are presented in Table 2.

2.3   Analyses

Assuming that those constructs most descriptive of a core endophenotype for marital status are also the most heritable, we used biometric models to estimate the genetic and environmental variance of the two proposed endophenotypes for marital status using 1972 and 1985 questionnaire response data. Because the variables were ordinal, we used PRELIS-2.12 (Joreskog and Sorbom, 1993b) to compute polychoric correlations of twins' marital status, assuming that ordinal measures reflected an underlying normal distribution with threshold values discriminating between categories. Biometric models were fitted to the polychoric correlation matrices using LISREL-8.12 (Joreskog & Sorbom, 1993a) weighted least squares analysis,

Table 1. Coding of Questionnaire Responses for successful pair bonding, 1972 and 1985

Successful heterosexual pair-bonding, 1972	N
1=Single	723
2=Divorced or Separated	738
3=Remarried	449
4=Married or Widowed	11,542

Successful heterosexual pair-bonding, 1985	N
1=Never-married	346
2=Multiple marriage, not married at present	122
3=Married once and terminated by divorce	267
4=Separated	48
5=Multiple marriage, married at present	892
6=Married once and married at present, or Married once and terminated by death	7,207

Table 2. Coding of marital status responses for marital instability,
1972 and 1985

Marital instability, 1972	N
1=Single	723
2=Married or Widowed	11,542
3=Divorced	449
4=Remarried	738

Marital Instability, 1985	N
1=Never-married	346
2=Married once and married at present, or Married once and terminated by death	7,207
3=Separated	48
4=Married once and terminated by divorce	267
5=Multiple marriage, married at present	892
6=Multiple marriage, not married at present	122

with asymptotic covariance matrices also created by PRELIS-2.12 (c.f., Neale and Cardon, 1992, on ordinal data analysis). Biometric models specified paths for contributions of additive genetic effects (a), genetic dominance (d), environmental effects common to both twins (c), and environmental effects not shared between twins (e). The latent variable representing genetic dominance (d) also reflects, mathematically, other non-additive genetic variance such as epistasis. Without identification of specific genetic loci, epistatic effects cannot be reliably measured. Similarly, the latent variable representing unique environment (e) also contained any error variance in the data. Models assumed additivity and orthogonality of components of variance.

Figure 1 shows the general structure of univariate analyses used in this study. Variances of latent genetic and environmental variables were fixed at 1.0. Additive and non-additive genetic variances were estimated, respectively, by squaring the values of the paths from the latent additive and non-additive genetic variables to the phenotype. Similarly, squaring the paths from latent common and unique environmental variables to the phenotype produced estimates of common and unique environmental variance. Models to be tested included triads of additive genetic, common environmental, and unique environmental variance (ACE) or of additive genetic, non-additive genetic, and unique environmental variance (ADE). Additive and non-additive genetic and common environmental variance were then each separately paired with non-shared environmental variance/error to form two-variable models, and finally, nonshared/error variance alone formed the reference model of no genetic or common environmental effects. Models' relative goodness-of-fit and parsimony were assessed using the Akaike Information Criterion (AIC) (Akaike, 1987), which balances goodness-of-fit and parsimony of the model by subtracting twice the model's degrees of freedom from its chi-square: AIC=Chi-square − 2df. The model with the lowest value for AIC was selected in each case as the best fitting and most parsimonious. Confidence intervals (95%) were also calculated for the models' parameters.[1]

3.   RESULTS

Despite selection effects, marital status of the NAS-NRC sample in 1972 and 1985 was comparable to that of its national birth cohort (U.S. Bureau of the Census, 1974, 1987). The percentage of married respondents in 1972 exceeded population percentages only slightly, and the percentage of widowed respondents was less than half that of the general population, although widowhood was infrequent (<1.5%) even in the national male

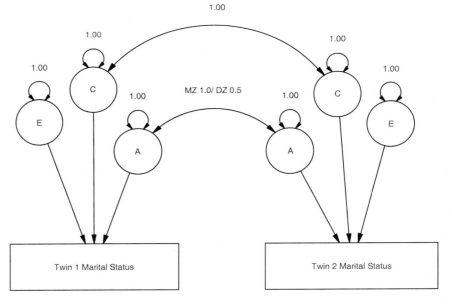

Figure 1: Biometric Model for Marital Status Endophenotype

cohort (U.S. Bureau of the Census, 1974). Although the percentage of widowed respondents in 1985 was comparable to U.S. population estimates for males then aged 55-64, fewer NAS-NRC respondents were divorced or single, and more were married than their counterparts in the general population (U.S. Bureau of the Census, 1987). Higher than expected rates of intact marriage in this sample may reflect response bias, demographic restrictions, and other selection factors. First, married twins from 1972 were more likely than their non-married counterparts to respond to the 1985 questionnaire. Second, NAS-NRC registry twins were all European-American, and in 1972 and 1985, divorce rates were lower for European-Americans than for African-Americans (U.S. Bureau of the Census, 1974, 1987). Because the U.S. Census (1974, 1985) did not report marital status rates by race within each cohort, we could not measure this ascertainment bias directly. Third, military selection for physical and mental health may account for higher rates of continuous marriage among this veteran sample.

Goodness-of-fit indices for all biometric models are presented in Table 3 with the best-fitting models highlighted in bold type. The parameters of best-fitting models for each endophenotype are presented in Table 4. The best-fitting models in all cases showed genetic (either additive or non-additive) and nonshared environmental contributions to endophenotypic variance. Confidence intervals around parameter estimates were sufficiently large that no significant differences in heritability emerged between our hypothesized endophenotypes for marital status. Nevertheless, the trends in the data were for broad heritability estimates to be higher for successful pair bonding than for frequency of marital status changes.

4. DISCUSSION

4.1 The question of endophenotype and declining heritability

Although the difference failed to reach statistical significance, the trend for pair bonding to be the more heritable endophenotype for marital status suggests that the tendency to form heterosexual pair bonds shows greater genetic variance than does the tendency toward having multiple mates. The pair bonding scale grouped divorce and never marrying more closely than did the other scale, and its greater genetic variance suggests that never marrying and divorce *sans* remarriage may be more similar to than different from one another, both endophenotypically and genetically.

Table 3: Goodness-of-fit indices for models of marital status endophenotype

Endophenotype	Model	χ^2	df	p	AIC
Pair bonding,	ace	0.28	0	1.00	0.28
1972	ade	0.00	0	1.00	0.00
	ae	**0.28**	**1**	**0.59**	**- 1.72**
	de	2.85	1	0.09	0.85
	ce	15.70	1	0.00	13.70
	e	135.53	2	0.00	131.53
Mate diversification,	ace	0.40	0	1.00	0.40
1972	ade	0.00	0	1.00	0.00
	ae	**0.40**	**0**	**0.53**	**- 1.60**
	de	0.61	1	0.44	- 1.39
	ce	6.27	1	0.01	4.27
	e	34.25	2	0.00	30.25
Pair bonding,	ace	0.33	0	1.00	0.33
1985	ade	0.00	0	1.00	0.00
	ae	0.64	1	0.42	-1.36
	de	**0.40**	**1**	**0.53**	**-1.60**
	ce	11.18	1	0.00	9.18
	e	73.04	2	0.00	69.04
Mate diversification,	ace	0.00	0	1.00	0.00
1985	ade	0.34	0	1.00	0.34
	ae	**0.34**	**1**	**0.56**	**-1.66**
	de	2.70	1	0.10	0.70
	ce	0.93	1	0.33	-1.07
	e	22.44	2	0.00	18.44

Table 4: Analyses of Marital Status Endophenotypes, 1972 and 1985 Questionnaires

Marital Status	Tetrachoric r's		Parameters of Best-Fitting Model			
	r_{MZ}	r_{DZ}	a^2	d^2	c^2	e^2
Pair bonding, 1972 (ae)	.424	.186	.42 (.23,.48)	--	-- (.00,.15)	.58 (.50,.66)
Pair bonding, 1985 (de)	.300	.113	-- (.00,.29)	.31 (.00,.41)	--	.69 (.60,.71)
Mate diversification, 1972 (ae)	.293	.112	.28 (.06,.37)	--	-- (.00, .16)	.72 (.62,.82)
mate diversification, 1985 (ae)	.209	.138	.22 (.00, .32)	--	-- (.00, .24)	.78 (.67,.90)

Both scales showed stronger genetic effects for the 1972 than for the 1985 ascertainment. Several explanations may account for this declining heritability over time. First, divorced and never-married men are at higher risk for mortality than the continuously married (Trovato, 1998), so disproportionate attrition of divorced or never-married individuals would lead to declining heritability. Additionally, as the veterans aged, they became increasingly more likely to become widowed, and the coding of "remarried" responses as previously divorced thereby introduced increasing error over time, which may contribute to a general decline in twin-co-twin correlations for divorce.

Second, there may be age-related changes in motivation to divorce so that, later in life, divorce is motivated less by heritable traits and more by environmental circumstances than it is earlier in life. Genetic variance in marital status may also decline over time because functions associated with marriage, procreation and nurturance of offspring lose their centrality as adults age. It may be that genetic influences associated with pair bonding are strongest at ages when the likelihood of procreation is greatest.

Third, secular divorce rates increased during the decade that elapsed between 1972 and 1985 and more widespread divorce may have contributed to declining heritability for divorce within the general population.

Nevertheless, the Minnesota data showed an opposite trend from the NAS-NRC data, as MZ-DZ differences in divorce correlation were larger among older (>40 years) than among younger twin pairs (McGue & Lykken, 1992). Unfortunately, age and cohort effects were confounded in both the Minnesota study and this one, as the Minnesota study examined different cohorts cross-sectionally and the current study followed a single cohort longitudinally. Distinctions between age and cohort effects will require following several cohorts longitudinally. At this time, it is impossible to discern the degree to which declining heritability of marital status reflects life course changes, societal changes, or sampling and ascertainment biases.

There is evidence of declining heritability over time in other traits associated with divorce risk. For example, both extraversion and neuroticism, traits associated with higher risk for divorce (Jockin et al., 1996), showed declining heritability with age (Viken, Rose, Kaprio, & Koskenvuo, 1994), and may account for some of the declining heritability of divorce. On the other hand, there is evidence that some personality traits show increasing heritability with age (Plomin & Nesselroade, 1990). Of course, sample attrition and fluctuating incidence of particular phenotypes affect any longitudinal study in population genetics, so findings of changing heritability in any phenotype over time must be interpreted with some caution.

From a slightly different perspective, fertility behavior has shown *increasing* heritability over generation cohorts that seems to reflect increasing options for individual choice, and hence, for the expression of individual differences (Kohler, Rodgers, & Christensen, 1999). This would suggest that, in the wake of late-twentieth century divorce reform in the United States, the heritability of marital status might be expected to increase. In addition, the passage of no-fault divorce law was followed by an increase in men's initiation and completion of divorce proceedings (Gunter & Johnson, 1978). This reversed the previous trend of women's initiation of the majority of divorce proceedings (Gunter & Johnson, 1978). The heritability of men's marital status particularly might be expected to increase over this period as men apparently increased their exercise of the option to divorce. The apparent decrease in heritability of marital status over time in the NAS-NRC sample is thus probably less likely due to societal changes and more likely due to sample attrition, measurement error, and the effects of age on the heritability of fertility-related behaviors.

4.2   Methodological concerns

There are obvious limitations to the inferences one can draw from a single-sex sample about genetic and environmental contributions to marital status. In a mixed sample, Jockin et al. (1996) found similar patterns across sexes in the heritability of divorce and its association with neuroticism, suggesting that some patterns of genetic and environmental contributions to divorce found in this all-male veteran sample may generalize across gender. Correlates of never-married status, however, differ by sex (Spreitzer & Riley, 1974). For example, at a population level, both IQ and psychosocial adjustment among never-married women tend to be higher than average, whereas those of never-married men tend to be lower than average (Spreitzer & Riley, 1974). Patterns of genetic and environmental influences on never marrying may be more sex-specific than those on divorce, and may therefore affect the heritability of pair bonding and of mate diversification. Additionally, evolutionary theory supports the idea of different optimal mating strategies for males and females, strategies whose genetic variability may therefore differ by sex.

4.3 Implications for further research

Obviously, marital status provides only an approximate description of heterosexual pair bonding and mate diversification patterns in contemporary versions of the human species. For example, never marrying does not necessarily preclude stable heterosexual pair bonding, as is evident in cultures where large numbers of couples live together in committed, but not legally binding, partnerships. Nor is stable pair-bonding limited to the heterosexual sphere, as homosexual partnerships can also reflect pair-bonding patterns. By contrast, the long duration of a legally contracted marriage does not necessarily signify a monogamous or successful partnership, as stable, unhappy marriages have been well-documented (Heaton & Albrecht, 1991; Levenson, Carstensen, & Gottman, 1993). Although divorce and never marrying may both be considered as "phenotypes," or observed characteristics of individuals at given times, they can also reflect a wide variety of behaviors. Greater precision in defining these behaviors may lead to a better understanding of how human pair bonds are formed and maintained.

In addition to psychopathology (Trumbetta et al., 1999) and personality (Jockin et al., 1996), other heritable characteristics likely affect population variability in marital status. For example, although limited by small and often self-selected samples, a growing volume of research suggests some genetic contribution to sexual orientation (Bailey & Martin, 1995; Bailey & Pillard, 1991; Eckert, Bouchard, Bohlen, & Heston, 1986; Hamer, Hu, Magnuson, Hu, & Pattatucci, 1993; See also Gladue, 1995 for a review), which may explain additional genetic variability in marital status. It would be difficult to study homosexuality in the NAS-NRC Twin Registry, given the social realities of the World War II veteran twin cohort and of the military during their time of service. It is very likely that recorded diagnoses of homosexuality in the VA records represent an underestimate of actual prevalence.

Another potential mediator of genetic effects on marital status is attachment security. There is evidence, for example, that marital satisfaction is higher for spouses with secure attachments than for those with insecure attachments (Feeney, 1996). There is also evidence that insecure, preoccupied, and disorganized attachment is more characteristic of violent than of non-violent husbands (Holtzworth-Munroe, Stuart, & Hutchinson, 1997). A recent, preliminary twin study of attachment suggests genetic contributions to attachment style (Finkel, Wille, & Matheny, 1998). Insofar as security of attachment influences the likelihood of marriage and divorce, and insofar as genetic factors contribute to attachment style, attachment may mediate genetic effects on marital status.

Animal studies have suggested additional pathways of potential genetic and environmental influences on pair bonding. For example, recent studies suggest neuroendocrinological bases for affiliative, pair bonding, and monogamous behaviors in non-human animals, and implicate oxytocin, vasopressin, and serotonin as potential mediators of these behaviors (Insel, 1997; Insel & Winslow, 1998; Insel, Young, & Wang, 1997; Young, Wang, & Insel, 1998). They may also be important mediators of genetic variability in behaviors associated with human pair bonding. From an environmental perspective, Ehrman's (Ehrman, 1990; Kim & Ehrman, 1998) *Drosophila* studies show the influence of environmental interventions at critical developmental stages on adult sexual behaviors, and suggest, by implication, that environmental influences at critical developmental stages may also influence human pair bonding behaviors. The exact nature of relevant environments and the identification of critical developmental stages for adult psychosexual behavior remain to be explored.

ACKNOWLEDGEMENTS

This research was supported in part by the U.S. Public Health Service research grant #RR11803 through the University of Louisville. We gratefully acknowledge the National Heart Blood Lung Institute for underwriting expenses of questionnaires 2 and 7, and all National Academy of Sciences staff who have labored to maintain this database. We are especially grateful to colleagues Eric Turkheimer of the University of Virginia and William F. Page of the National Academy of Sciences for their contributions, as well as to the participants in the World War II Veteran Twin Registry for their generous cooperation over many years. The opinions and assertions contained herein are those of the authors and are not to be construed as reflecting the views or position of the anational Academy of Sciences, the Institute of Medicine, or the National Research Council.

REFERENCES

Akaike, H. (1987). Factor analysis and AIC. *Psychometrika, 52*, 317-332.

Bailey, J.M., & Martin, N.G. (1995). An Australian twin study of sexual orientation. *Behavior Genetics, 25*, 254.

Bailey, J.M., & Pillard, R.C. (1991). A genetic study of male sexual orientation. *Archives of General Psychiatry, 48*, 1089-1096.

Carmelli, D., Swan, G.E., & Robinette, D. (1991). Substance use in World War II veteran twins: A genetic analysis. *Epidemiology in Military and Veteran Populations.* Washington, D.C.: National Academy Press.

Dunbar, R. I. M. (1995). The mating system of callitrichid primates: I. Conditions for the coevolution of pair-bonding and twinning. *Animal Behaviour, 50*, 1057-1070

Eckert, E.D., Bouchard, T.J., Bohlen, J., & Heston, L.L. (1986). Homosexuality in monozygotic twins reared apart. *British Journal of Psychiatry, 148*, 421-425.

Ehrman, L. (1990). Developmental isolation and subsequent adult behavior of Drosophila pseudoobscura. *Behavior Genetics, 20*, 609-615

Fabsitz, R.R., Kalousdian, S., Carmelli, D., Robinette, D., & Christian, J.C. (1988). Characteristics of participants and nonparticipants in the NHBLI twin study. *Acta Geneticae Medicae Gemellologiae, 37*, 217-228.

Feeney, J.A. (1996). Attachment, caregiving, and marital satisfaction. *Personal Relationships, 3,* 401-416.

Finkel, D., Wille, D.E., & Matheny, A.P. (1998). Preliminary results from a twin study of infant-caregiver attachment. *Behavior Genetics, 28,* 1-8.

Gladue, B.A. (1995). The biopsychology of sexual orientation. *Current Directions in Psychological Science, 3*, 150-154.

Gunter, B.G., & Johnson, D.P. (1978). Divorce filing as role behavior: Effect of no-fault law on divorce filing patterns. *Journal of Marriage & the Family, 40*, 571-574.

Hamer, D.H., Hu, S., Magnuson, V.L., Hu, N., & Pattatucci, A.M.L. (1993). A linkage between DNA markers on the X chromosome and male sexual orientation. *Science, 261,* 321-327

Heaton, T.B., & Albrecht, S.L. (1991). Stable unhappy marriages. *Journal of Marriage and the Family, 53*, 747-758.

Holtzworth-Munroe, A., Stuart, G.L., & Hutchinson, G. (1997). Violent versus nonviolent husbands: Differences in attachment patterns, dependency, and jealousy. *Journal of Family Psychology, 11,* 314-331.

Hrubec, Z. (1973). The effect of diagnostic ascertainment in twins on the assessment of the genetic factor in disease etiology. *American Journal of Human Genetics, 25,* 15-28.

Insel, T.R. (1997). A neurobiological basis of social attachment. *American Journal of Psychiatry, 154,* 726-735.

Insel, T.R., & Winslow, J.T. (1998). Serotonin and neuropeptides in affiliative behaviors. *Biological Psychiatry, 44,* 207-219.

Insel, T.R., Young, l., & Wang, Z. (1997). Molecular aspects of monogamy. *Annals of the New York Academy of Sciences, 807,* 302-316.

Jablon, S., Neel, J.V., Gershowitz, H., & Atkinson, G.F. (1967). The NAS-NRC Twin Panel: Methods of construction of the panel, zygosity diagnosis, and proposed use. *American Journal of Human Genetics, 19,* 133-161.

Jockin, V., McGue, M., & Lykken, D.T. (1996). Personality and divorce: a genetic analysis. *Journal of Personality & Social Psychology, 71,* 288-99.

Joreskog, K.G., & Sorbom, D. (1993a). *LISREL8: User's Reference Guide.* Chicago, Illinois: Scientific Software International, Inc.

Joreskog, K.G., & Sorbom, D. (1993b). *PRELIS2: User's Reference Guide.* Chicago: Scientific Software International, Inc.

Kelly, E.L., & Conley, J.J. (1987). Personality and compatibility: A prospective analysis of marital stability and marital satisfaction. *Journal of Personality and Social Psychology, 52,* 27-40.

Kendler, K.S. (1986). A twin study of mortality in schizophrenia and neurosis. *Archives of General Psychiatry, 43,* 643-649.

Kendler, K.S., & Robinette, C.D. (1983). Schizophrenia in the National Academy of Sciences-National Research Council Twin Registry: A 16-year update. *American Journal of Psychiatry, 140,* 1551-1563.

Kim, Y-K., & Ehrman, L. (1998). Developmental isolation and subsequent adult behavior of Drosophila paulistorum. IV. Courtship. *Behavior Genetics, 28,* 57-65.

Kohler, H-P, Rodgers, J.L., & Christensen, K. (1999). Is fertility behavior in our genes? Findings from a Danish twin study. *Population and Development Review, 25,* 253-288.

Komers, P.E. (1996). Obligate monogamy without paternal care in Kirk's dikdik. *Animal Behaviour, 51,* 131-140.

Larsen, T. (1991). Anti-predator behaviour and mating systems in waders: Aggressive nest defence selects for monogamy. *Animal Behaviour, 41,* 1057-1062.

LeClere, F.B., Rogers, R.G., Peters, K.D. (1997). Ethnicity and mortality in the United States: Individual and community correlates. *Social Forces, 76,* 169-198

Levenson, R.W., Castensen, L.L., & Gottman, J.M. (1993). Long-term marriage: Age, gender, and satisfaction. *Psychology and Aging, 8),* 301-313.

McGue, M., & Lykken, D.T. (1992). Genetic influence on the risk of divorce. *Psychological Science, 3,* 368-373.

McGuffin, P., Owen, M., O'Donovan, M., Thapar, A., & Gottesman, I.I. (1994). *Seminars in Psychiatric Genetics.* Washington DC: American Psychiatric Press.

Morell, V. (1998). A new look at monogamy. *Science, 281,* 1982-1983.

Neale, M.C., & Cardon, L.R. (1992). *Methodology for Genetic Studies of Twins and Families.* Boston: Kluwer Academic Publishers.

Neale, M.C. Eaves, L.J., & Kendler, K.S. (1994). The power of the classical twin study to resolve variation in threshold traits. *Behavior Genetics, 24,* 239-258.

Piper, W.H., Evers, D.C., Meyer, M.W., Tischler, K.B., Kaplan, J.D., & fleischer, R.C. (1997). Genetic monogamy in the common loon (*Gavia immer*). *Behavioral Ecology & Sociobiology, 41,* 25-31.

Plomin, R., & Nesselroade, J.R. (1990). Behavioral genetics and personality change. *Journal of Personality, 58,* 191-220.

Salo, A.L., Shapiro, L.E., & Dewsbury, D.A. (1993). Affiliative behavior in different species of voles (Microtus*). Psychological Reports, 72,* 316-318.

Spreitzer, E., & Riley, L.E. (1974). Factors associated with singlehood. *Journal of Marriage and the Family, 36,* 533-542.

Trovato, F. (1998). Nativity, marital status and mortality in Canada. *Canadian Review of Sociology & Anthropology, 35,* 65-91.

Trumbetta, S.L., Gottesman, I.I., Turkheimer, E.N., & Page, W.F. (1999). Genes for marital status? Divorce, never marrying, and psychopathology in the NAS-NRC World War II Veteran Twin Registry. Manuscript submitted for publication.

U.S. Bureau of the Census (1974). *Statistical Abstract of the United States: 1974.* Washington, D.C.: U.S. Government Printing Office.

U.S. Bureau of the Census (1987*). Statistical Abstract of the United States: 1987.* Washington, D.C.: U.S. Government Printing Office.

VanSchaik, C.P., & Dunbar, R.I. (1990). The evolution of monogamy in large primates: A new hypothesis and some crucial tests. *Behaviour, 115,* 30-62.

Viken, R.J., Rose, R.J., Kaprio, J., & Koskenvuo, M. (1994). A developmental genetic analysis of adult personality: Extraversion and neuroticism from 18 to 59 years of age. *Journal of Personality and Social Psychology, 66,* 722-730.

Young, L.J., Wang. Z., & Insel, T.R. (1998). Neuroendocrine bases of monogamy. *Trends in Neurosciences, 21,* 71-75.

Notes

[1] Because LISREL 8.12 could not directly produce accurate confidence intervals for parameters in these analyses, confidence intervals were calculated in the following way. Each path of the biometric model (a,c,e) was fixed, iteratively, to a series of values ascending and descending from its estimated value in the best-fitting model, while the other two paths in the model were free to vary. Because fixing the value added one degree of freedom to the model, upper and lower threshold estimates producing, at one degree of freedom, a significant chi-square change from that of the best-fitting model represented the upper and lower bounds, respectively, of that path's 95% confidence interval. Squaring these path values (a,c,e) produced the upper and lower bounds of the 95% confidence interval for the corresponding variance parameter (h^2, c^2, e^2).

Chapter 12

AGE AT FIRST CHILDBIRTH: A MAJOR SELECTIVE FACTOR FOR PSYCHIATRIC GENES IN THE TWENTIETH CENTURY

David E. Comings

Key words: gene selection, fertility, behavior

Abstract: In many different epidemiological studies using structured psychiatric instruments, when the data are analyzed by age cohorts there is a significant trend for the frequency of disorders to increase, and for the age of onset to decrease, in younger cohorts. This trend, occurring over the past 60 years, has been true of a wide range of psychiatric disorders. Twin and adoption studies have shown that each of these disorders has a significant genetic component, due to the additive and interactive effect of multiple genes. While the authors have not been able to identify the reasons for these trends, it has generally been assumed that the changes are occurring too rapidly to be genetically based. However, one of the expectations of polygenic disorders is that as the genetic loading increases, the frequency of the disorder increases and the age of onset decreases. I propose that there is a factor, new to the 20th century, that could account for changes in the gene pool of this speed and magnitude, and that is the rapid increase in the number of individuals receiving a higher education. Percent of individuals in the U.S. attending college has increased from 2 percent in 1920 to 37 percent in 1980. The reason this can be a strong genetic selective factor is two-fold. First, there is a high correlation between age of onset of child bearing and years of education. Those who drop out of school before completing high school initiate child bearing between 20 and 23 years of age. By contrast, those who attend college and/or graduate school initiate child bearing between 26 to 28 or more years of age, and have fewer children. Second, the frequency of learning, addictive and other behavioral disorders is higher in those who drop out of school early compared to those who remain in school. As a result, the genes involved in these disorders turn over more rapidly than the genes of those whole remain in school. To verify or refute this hypothesis, there is a need for a new field, *molecular demography*, to monitor the changes in the frequency of genes across age groups.

1.   INCREASES IN BEHAVIORAL DISORDERS

In 1978 the President's commission on psychiatry initiated the funding of a project to develop an instrument suitable for epidemiological studies of the prevalence of psychiatric disorders in the United States. This was developed by Lee Robins at the University of Washington in St. Louis and called the Diagnostic Interview Schedule (DIS). This structured instrument allowed the launching of the Epidemoligic Catchment Area (ECA) study of the prevalence of psychiatric disorders in over 18,000 citizens in five U.S. cities and provided a valuable tool for researchers interested in the genetics of these disorders.

In 1985 Klerman et al (1985) used the DIS to examine the frequency of major depression in the relatives of patients that had major depression themselves. Rather than simply report on the lifetime prevalence of major depression in all relatives, Klerman et al divided them into six birth cohorts consisting of those born before 1910 and in five subsequent 10-year intervals. Since the rates of lifetime depression were being examined one would anticipate that the prevalence would be greatest for those who lived the longest. However, just the opposite was observed. There was a dramatic and significant increase in the prevalence of major depression progressing

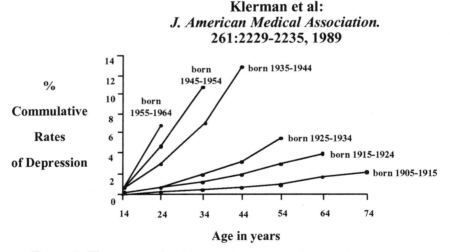

Figure 1. The progressive increase in the prevalence and decrease in age of onset in major depression by age cohort. Redrawn (Comings, 1996) from Klerman et al (1989) by permission, J. Am. Med. Assn

from the older to younger cohorts. In addition, the age of onset of major depression progressively decreased in each of the younger cohorts. Since depression is more common in women, it was not surprising that the prevalences for all cohorts were higher in female compared to male relatives. To examine the possibility that this was simply an artifact of examining the relatives of patients with depression, in 1989 Klerman and Weissman (1989) examined the prevalence of lifetime depression by age cohort in the general population. Figure 1 illustrates these results. As with the earlier study, again there was a dramatic increase in the prevalence of major depression and an earlier age of onset for the progressively younger cohorts.

To determine if these trends could be generalized to other cities and other countries, an extensive cross-national study of major depression was launched. This showed that the same trends were present in other cities in the U.S. and in Canada, France, Germany, Lebanon, New Zealand, and Taiwan (Cross-National Collaborative Group, 1992). The major exceptions were in less developed countries like Puerto Rico, where the rates of depression showed a similar level in all age cohorts. To determine if the results were unique to major depression, Gershon et al (1987) examined two other mood disorders, schizoaffective disorder and manic depressive disorder. The same age cohort trends were present. This was important because while one could blame major depression on an increasingly complex society, schizoaffective disorder and manic-depressive disorder have a strong biological and genetic component. An environment predisposing to depression should not cause mania.

Even more remarkably, virtually every behavioral disorder examined using structured assessments, including alcoholism, drug abuse, anxiety, obsessive compulsive disorder, phobias, schizophrenia, autism, leaning disabilities and disruptive childhood disorders, have shown the same trends (Comings, 1996). One of the most dramatic changes has been the increase in learning disabilities (Figure 2). Based on the Annual Report to the Congress, the number of children requiring special education for learning disabilities increased form 890,000 in 1976 to 2,714,000 in 1993, an increase that far outpaced the increase in population or increase in number of children in school. In most school districts, getting a child in such classes is not easy and requires extensive evaluation by one or more educational psychologists and often requires an individual education plan and legal help.

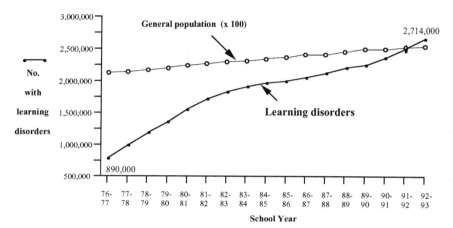

Figure 2. Progressive increase in the frequency of students in special classes for learning disabilities in the United States, from 1976 to 1993. From Comings, D. E. *The Gene Bomb*. Hope Press, Duarte, CA, 1996 by permission

While there have been many attempts to explain these trends, including socioeconomic, environmental and other factors (Gershon et. al., 1987; Lavori et. al., 1987; Simon and VonKorff, 1992; Simon et. al., 1995) none has been satisfactory (Comings, 1996). Poor memory in the older cohorts, or a tendency to be less open to an interview that is based on being self-revealing, or simply being less attuned to the psychological processes being probed, could produce the same trends. There are a number of results that make these explanations less likely. First, in the studies of depression in cohorts, done in countries like Puerto Rico (Cross-National Collaborative Group, 1992) where higher education limited, showed the same frequency of depression in the older as the younger cohorts. It is unlikely that these individuals have a better memory for these events than older subjects in industrially advanced countries. Second, studies of disorders that individuals are unlikely to forget, such as alcohol and drug dependence, manic-depressive and schizoaffective disorder showed the same cohort effect (Gershon et. al., 1987; Cloninger et. al., 1988; Dinwiddie and Reich, 1991; Simon and VonKorff, 1992). Finally, when the studies were limited only to recalling long episodes of severe depression and limited to subjects 50 years of age or less, where poor memory and other factors would be much less likely, the results were the same (Lavori et. al., 1987). Since all of these disorders have a strong genetic component the possibility of genetic changes has also been considered and rejected (Gershon et. al., 1987) on the assumption that the changes, occurring over a period of one century, have been too rapid.

2.   HIGHER EDUCATION AND AGE OF FIRST CHILDBIRTH AS A FORCE FOR GENE SELECTION

In this chapter, and previously (Comings, 1996), I re-examine the possibility that rapid changes in the frequency of a group of behavior genes is, in fact, a viable explanation for these trends. While there have been many changes both technological and social in the past century, most of these do not necessarily have a strong inherent ability to be a significant force in gene selection. One important exception is the dramatic increase in the past century, in this country and throughout the industrialized world, in the number of citizens that have attended college. In the United States this has increased from 2 percent in 1920 to over 37% in 1990 (Herrnstein and Murray, 1994). The reason this can be a selective force is that data from the National Longitudinal Survey of Youth show there is a dramatic correlation between years of education and age of initiation of childbirth (Comings, 1996). As shown in Figure 3, those subjects that drop out before graduating from High School tend to initiate childbearing between 20.5 to 22.8 years of age. By contrast, the age at the birth of the first child averages 24 years for those with two years of college education, 26.5 years for those completing a college education, and 27.5 years for those doing graduate work. The correlation coefficient, r, between age at the birth of the first child and years of education was .39, $r^2 = .15$, p $< .0000001$. This is understandable since those who go on to college and possibly graduate school, usually postpone having children until their education is completed, and until they are married, have a job, and feel financially able to start a family. As shown, the age at first birth in this group is often delayed to between 26 and 30+ years of age compared to approximately 21 years of age for those who do not finish high school. The difference in age between the two groups average 6 years. Since IQ correlates very highly with years of education, it should also

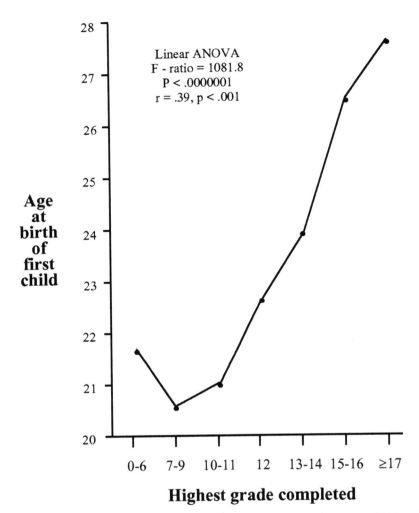

Linear ANOVA
F - ratio = 1081.8
P < .0000001
r = .39, p < .001

Highest grade completed

Figure 3. The correlation between age at the birth of the first child and highest grade completed, from NLSY database. From Comings, D. E. *The Gene Bomb.* Hope Press, Duarte, CA, 1996 by permission

be highly correlated with the age of initiation of childbirth. As shown in Figure 4, in the NLSY data there was a highly significant correlation between IQ percentile and average at the birth of the first child. The correlation coefficient was over .7 and r^2 was over .5.

In-and-of-itself, these findings would not necessarily suggest a selective factor for behavioral genes. However, if those individuals having children

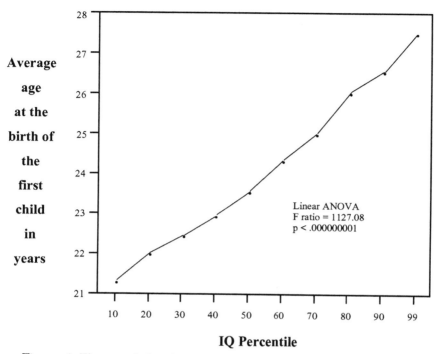

Figure 4. The correlation between age at the birth of the first child and IQ, from NLSY database. From Comings, D. E. *The Gene Bomb*. Hope Press, Duarte, CA, 1996 by permission

early tend to have different behaviors than those having children late, and if those behaviors have a significant genetic component, because of the more rapid turnover of the genes associated with early ages of childbirth there would be a selection for those genes. Ninety percent of mothers have their children between ages 18 and 30 years of age, over a span of 12 years (Comings, 1996). A difference of 6 years between those having children early versus those having children late is 50 percent of this span and as such the age differential in childbirth can be a powerful selective force. As the average number of children per family decreases, this can be a much more significant factor than the number of children an individual has.

A study by Maria Kovacs et al (1994) from the University of Pittsburgh School of Medicine illustrates the principle. She was interested in the question of the causes of teenage pregnancy. She initiated a longitudinal study of 83 girls between age 8 and 13 all of a low but similar socioeconomic status. Her *a priori* hypothesis was that those girls with depression would be the ones susceptible to teenage pregnancy. To test this she performed psychological testing and psychiatric diagnosis at the above ages then followed the girls longitudinally until they were 20 years of age. To her surprise, there was no correlation between depression and teenage pregnancy. The overwhelming factor was the presence or absence of conduct disorder. Figure 5 shows these results. The mean age of first pregnancy for the 33 girls with a diagnosis of conduct disorder was 17 years. Since the cumulative incidence of pregnancy had not quite reached 50% for the non-conduct disorders girls, for comparison purposes this curve was extrapolated from 20 to 25 years of age. Thus, by contrast, the mean age of first pregnancy for the 50 girls without conduct disorder was 24 years, a

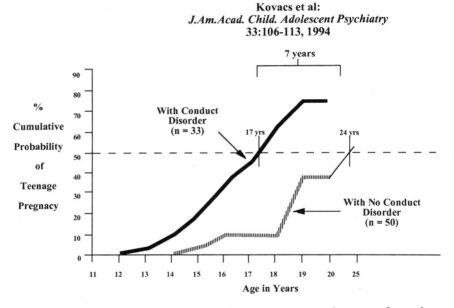

Figure 5. Association between the presence or absence of conduct disorder and probability of teenage or later pregnancy. Extrapolated from Kovacs et al (Kovacs et. al., 1994). From Comings, D. E. the Gene Bomb. Hope Press, Duarte, CA, 1996 by permission.

difference of 7 years. In addition to showing the important role of conduct disorder in teenage pregnancy, this study was also important because it showed that socioeconomic status was not a factor. Twin studies of conduct disorder have shown that genetic factors account for 71 to 90 percent of the variance (Slutske et. al., 1997), indicating conduct disorder is predominately a genetic disorder. It is possible to obtain a rough estimate of the relative selective advantage provided by conduct disorder by assuming that the earlier the initiation of childbearing the more rapidly that person's genes turnover. Thus, on average the genes in these girls with conduct disorder would turn over about 29 percent more rapidly (7/24.0 = .29). Another way to think of it is that assuming the presence or absence of conduct disorder was the same in subsequent generations, and each had only one child, the girls with conduct disorder would have 100/17 or 5.88 generations per 100 years, while the non-conduct disorder girls would have 100/24.0 or 4.17 generations per 100 years. This is a (5.88-4.17)/5.88 or .29 or 29 percent

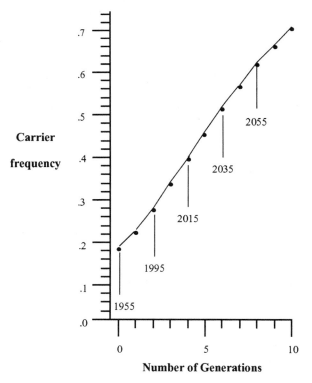

Figure 6. Progressive increase in the frequency of an allele with a 1.25 fold selective advantage, over numerous generations

selective advantage for the conduct disorder genes. When this degree of selection is used in equations that show the rate of increase in the frequency of single alleles, it shows their frequency could almost double over a half-century. This is illustrated in Figure 6 using the formula of Falconer (1981) for calculating the rate at which an allele of a gene that has a selective advantage over the other allele(s) increases in frequency over successive generations. In this illustration I have assumed that the allele being selected has a 1.25 fold selective advantage over the other allele(s). This shows that if we assume that the frequency of a given allele is .18 in 1955, and each generation is 20 years, by 1995 the frequency of that allele would increase to .28, and by 2015 to .40, more than doubling over a period of 60 years.

There are many caveats that apply to this illustration. The most important assumption is that behavioral disorders are polygenic rather than due to single mutant genes. This complicates the process because there are presently no studies of the rates of selection of genes in human polygenic disorders. However, whether the actual rate of selection of the genes for polygenic disorders is lower or higher than the assumed rate, the principle is still valid. Given a selective advantage in the range of 1.1 to 1.3, whether the time period is 4 or 8 generations, and whether 4 or 10 or more genes are involved, the frequency of the involved alleles will progressively increase. Figure 7 summarizes a large amount of data derived from three different data bases in two different countries, the NLSY and the National Center for Health Statistics in the United States, and the National Child Development Study in England (Comings, 1996). These data show the trends in behavioral variables between individuals having children early versus those having children late. Based on the NLSY data there is a high correlation between the age at the birth of the first child and IQ (r = .40, p < .000001) and the number of children (Rodgers et. al., 2000) (r = .99). All the addictive behaviors were significantly more common in those having children earlier.

Of all the behaviors the correlation with learning disorders is probably the easiest to understand. Most children with learning disorders have great difficulty with school, poor self-esteem and tend to drop out of school early and start a family, with or without getting married. By contrast, those without learning disabilities are more likely to enjoy school, go to college and delay childbearing until their late 20's and 30's. While I am aware of only one twin study of learning disabilities, that showed higher concordance for identical twins (Lewis and Thompson, 1992) consistent with clinical experience that genetic factors play a significant role. The disparity in the age of initiation of childbirth in those with learning disabilities would select for the genes involved.

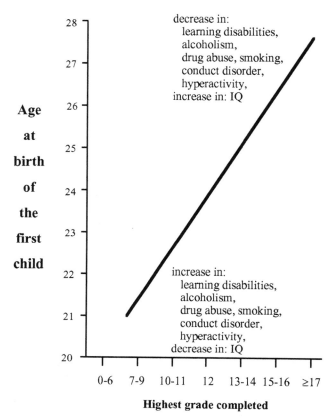

Figure 7. Summary of information from three databases in two countries on the association between addictive, disruptive, and learning disorders and IQ by age at the birth of the first child

As discussed above, the two critical factors occurring in the 20th century, contributing to the potential for increased selection for these genes, are the increase in the number of individuals, especially women, attending college, and the delay in child bearing that a college education produces. The comparison of the product of these two factors in 1920 versus 1947 versus 1965 versus 1978 is shown in Figure 8. Figure 8A shows that the percent of women attending college increased from 2 to 12 to 34 to 36 across these intervals. Figure 8B shows that the delay in child bearing that a college and postgraduate education produces increased from 4.0 to 8.9 years over this time period. The product of these two (Figure 8C) indicates the combined effect of these two factors. This increased from 8 in 1920 to 322 in 1978, a 40-fold increase in the effect of higher education as a selective factor. Even

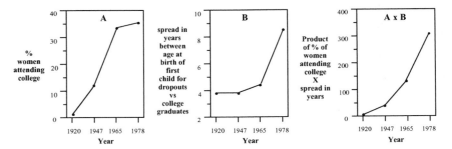

Figure 8. Increase in percent of women attending college (A), difference in ages of first child birth in high school drop-outs vs. college students (B), and the product of A and B (C). From Comings, D. E. The Gene Bomb. Hope Press, Duarte, CA, 1996 by permission

if some of the women shown in Figure 8A did not complete college, the product would still approach a 30-fold increase. While the science of understanding the selection of alleles in polygenic disorders is in its infancy, my suspicion is that this type of change, especially when it directly affects reproductive practices, is sufficient to account for the observed cohort changes in the frequency of impulsive, compulsive, addictive and other behaviors.

It is reasonable to ask if there are there any other factors, in addition to the role of higher education and genetic factors, as an explanation for changes in the age of onset and frequency of behavioral disorders? In this regard, is of interest to note that Gershon et al (1987) commented, "there were no obvious social or historical events or trend" that could readily explain this data. While the rise in the use of contraception is one reasonable candidate, I would argue that it was the availability of contraception that allowed college and post-graduate students to postpone childbearing and was responsible for the dramatic increase between 1965 and 1978 in spread in years between the age at the birth of the first child for those who dropped out of school versus those attending college, shown in Figure 8B. Thus, contraception was a factor but it had to be coupled with higher education to have an effect on gene selection. While the rise of feminism could have played a role, I would argue that feminism and the attainment of a higher education are closely related factors in postponing childbearing. However, the number of women who attain a higher education is far greater than the number who are also feminists. While the size of families has dropped considerably in the past century, this alone is not a factor in gene selection. It must be coupled with a differential drop in family size for those with or without these behavioral disorders. Thus, in addition to a delay in onset of childbearing, the NLSY data show that higher education is also associated with a decrease in the number of children couples have. A final factor, is the dramatic rise in the number of unwed mothers in the 20th century (Herrnstein and Murray, 1994). However, this is also subsumed under the role of education since teenage pregnancy is one of the causes of dropping out of school.

3. INDIVIDUAL GENES

3.1 DRD2 Gene

The study and identification of individual genes involved in various behavioral disorders has only occurred to any significant degree in the past 8 years – since the development of the polymerase chain reaction (PCR). One of the most studied genes is the dopamine D_2 receptor gene *(DRD2)*. This gene has been implicated in a range of impulsive, compulsive, addictive behaviors including alcoholism, drug dependence, smoking, pathological gambling, compulsive eating, ADHD, Tourette syndrome, conduct disorder and posttraumatic stress disorder (Blum et. al., 1995, 1996; Comings et. al., 1991, 1996). Table 1 shows the results of our examination of the prevalence of the *Taq* I A1 allele of the *DRD2* gene in students at the California State University at San Bernardino with no selection to exclude substance abuse. While the prevalence of the A1 allele (11 + 12 genotype) for the whole group was 33 percent, it was 40.5 percent for those less than 35 years of age versus 26.6 percent for those greater than 35 years of age. In addition, in the less than 35 years of age group, those who carried the A1 allele had an average of .91 children while those who did not carry this allele had an average of only .43 children. By contrast, of those over age 35 years of age, those that carried the A1 allele had an average of 1.28 children while those who did not carry this allele had an average of 1.71 children. A hierarchical ANOVA shows that as expected there was a significant correlation between age and the number of children. For the total group there was no association between the number of children and the A1 allele (p =.823). However,

---

Table 1. DRD2 Taq I A1/A2 Alleles and Age of Childbearing: Number of Children

---

	Means (n)			
	D_2A 11 or 12	D_2A 22	Total	% D_2A 11 or 12
<35 yrs	.91 (47)	.43 (59)	.63 (116)	40.5
≥35 years	1.28 (36)	1.71 (99)	1.59 (135)	26.6
Total	1.07 (83)	1.18 (168)	1.15 (251)	33.0

Hierarchial ANOVA

Main effects	F-ratio	p
Age (<35/≥35)	36.08	.000
DRD2	0.80	.823
Total	18.06	.000
2-Way interaction		
Age-*DRD2*	6.99	.009

---

there was a significant 2-way interaction of Age x *DRD2*, reflecting the fact that in the younger subjects having children was associated with the A1 allele, while in the older group, having children was associated with the absence of the A1 allele. These results are consistent with the concern that the frequency of certain alleles of genes associated with several behavioral disorders are more common in individuals in younger cohorts.

At the opposite end of the age spectrum, Noble et al (1997) examined the prevalence of the D_2A1 allele in the older population examining the hypothesis that it might be associated with Alzheimer's disease. It was not. Instead, an analyses by older age cohorts, including both Alzheimer's disease patients and controls, indicated there was a decrease in the frequency of the A1 allele frequency with increasing age. While they suggested this might be due to an association between the A1 allele and diseases causing early death, it is equally possible that this is simply an extension into the older age group of a trend for an increase in the frequency of the A1 allele in younger cohorts.

3.2   NOS3 Gene

The results for the endothelial nitric oxide synthase gene are presented in the chapter by MacMurray et al (1999) in this volume. In a series of parents of twins, the 2 allele of the single base pair nucleotide polymorphism of the endothelial *NOS3* gene (Wang et. al., 1996) was associated with increased monozygotic twinning, decreased age of the mother at the birth of the twins, decreased amount of higher education, and higher aggression scores based on the Multidimensional Personality Questionnaire. The 2 allele was present at higher frequency in younger subjects.

The results for both of these genes, showing higher frequencies in younger subjects of alleles associated with addictive, compulsive and aggressive behaviors, is consistent with the progressive selection of genes associated with impulsive, compulsive, and addictive behaviors. It is also consistent with the hypothesis that the age of first birth of children is a factor in this selection.

4.   THE NEED TO LAUNCH A NEW FIELD: MOLECULAR DEMOGRAPHY

There is sufficient evidence from epidemiology, psychiatry, demography, and molecular genetics to suggest that the progressive increase in the frequency of a range of behavioral disorders is real and largely driven by gene selection. I believe the only new factor developing in the 20th century with sufficient effect on reproductive performance to account for these effects is the increase in higher education. As discussed previously (Comings, 1996), I am pessimistic that any social changes will have the power to change these trends. I believe the first thing that needs to be done is to further verify, at a molecular genetic level, if this selection is occurring. The simplest way to do this is to launch a new field of investigation that I call *molecular demography* – the study of the frequency of genes in different age cohorts from newborns to 80+ years of age. Changes in the frequency of genes at the elder end of life would also identify genes important in longevity. The genotyping of several dozen genes known or suspected of being associated with impulsive, compulsive, addictive behaviors and learning disorders should be performed. Such a relatively modest study would definitively answer the question of whether gene selection is taking place and whether it is a continuing or accelerating process.

My experience with discussing these issues with many audiences is that there is almost universal concern, but it is difficult to get very excited about something that progresses so slowly that its effects are only apparent when the trends over many years and several generations are examined. It is also difficult be get excited about something that has not been proven to be true. If the predicted changes in the frequency of alleles associated with disruptive and addictive behaviors are verified, the implications for the future mental health of society are profound.

REFERENCES

Blum, K., Cull, J. G., Braverman, E. R., Comings, D. E. (1996) Reward deficiency syndrome. *American Scientist , 84*, 132-145.

Blum, K., Wood, R. C., Sheridan, P. J., Chen, T., Comings, D.E. (1995) Dopamine D2 receptor gene variants: Association and linkage studies in impulsive, addictive and compulsive disorders. *Pharmacogenetics, 5,* 121-141.

Cloninger, C. R., Reich, T., Sigvardsson, S., von Knorring, A. L., Bohman, M. (1988) Effects of changes in alcohol use between generations on the inheritance of alcohol abuse. In R. M. Rose & J. E. Barrett (Eds), *Alcoholism: Origins and Outcome*. New York: Raven Press.

Comings, D. E. (1996). *The Gene Bomb: Does Higher Education and Advanced Technology Accelerate the Selection of Genes for Learning Disorders, ADHD, Addictive, and Disruptive Behaviors?* Duarte, CA: Hope Press.

Comings, D. E., Comings, B. G., Muhleman, D., Dietz, G., Shahbahrami, B., Tast, D., Knell, E., et al. (1991). The dopamine D2 receptor locus as a modifying gene in neuropsychiatric disorders. *Journal of the American Medical Association, 266*,1793-1800.

Comings, D. E., Rosenthal, R. J., Lesieur, H. R., Rugle, L., Muhleman, D., Chiu,C., Dietz, G., et al: (1996). A study of the dopamine D2 receptor gene in pathological gambling, *Pharmacogenetics, 6*, 223-234.

Cross-National Collaborative Group. (1992). The changing rate of major depression. *Journal of the American .Medical.Association. 268* , 3098-3105.

Dinwiddie, S.H., Reich, T. (1991). Epidemiological perspectives on children of alcoholics. *Recent.Dev.Alcohol, 9,* 287-299.

Falconer, D. S. (1981*). Introduction to Quantative Genetics.* New York: Longman.

Gershon, E. S., Hamovit, J. H., Guroff, J. J. (1987). Birth-cohort changes in manic and depressive disorders in relatives of bipolar and schizoafective patients. *Archives of General Psychiatry, 44,* 314-319.

Hermstein,R.., Murray, C. (1994). *The Bell Curve.* New York: The Free Press.

Klerman, G. L., Lavori, P. W., Rice, J., et.al. (1985). Birth-cohort trends in rates of major depressive disorder among relatives of patients with affective disorder. *Archives of GeneralPsychiatry 42,* 689-693.

Klerman, G. L., Weissman, M.M. (1989). Increasing rates of depression. *Journal of the American Medical Association, 261,* 2229-2235.

Kovacs, M., Krol, R. S. M., Voti, L. (1994). Early onset psychopathology and the risk of teenage pregnancy among clinically referred girls. *Journal of the American Academy of Child and Adolescent Psychiatry , 33,* 106-113.

Lavori, P. W., Klerman, G. L., Keller, M. B., Reich, T., Rice, J., Endicott, J. (1987). Age-period-cohort analysis of secular trends in onset of major depression: Findings in siblings of patients with major affective disorder. *Journal of Psychiatry Research, 21,* 23-35.

Lewis, B. A. & Thompson, L. A. (1992). A study of developmental speech and language disorders in twins. *Journal of .Speech and .Hearing Research, 35,* 1086-1094.

MacMurray, J.P., Kovacs, B., McGue, M., Legro, R., Blake, H., Comings, D.E. (1999). Associations between the entothelial nitric acid synthase gene (NOS), reproductive behaviors, and twinning. In J. L. Rodgers, D. C. Rowe, W. B. Miller (Eds) *Genetic Influences on Human Fertility and Sexuality,* Mahweh, MA: Kluwer.

Noble, E. P., Matsuyama, S. S., Jarvik, L. F., Komo, S., Kaplan, A., Ritchie, T., Pritchard M.L., et al. (1997). D2 dopamine receptor A1 allele in Alzheimer disease and aging. *Archives of Neurology , 54,* 281-285.

Rodgers, J. L., Cleveland, H. H., van den Ord, E., Rowe D. C. (2000). Do large families make low-IQ children, or do low-IQ parents make large families? *American Psychologist,* in press.

Simon, G.E., VonKorff, M. (1992) Reevaluation of secular trends in depression rates. *Archives of General Psychiatry , 37* ,511-513.

Simon, G.E., VonKorff, M., Ustun, T.B., Gater, R., Gureje, O., Sartorius, N. (1995). Is the lifetime risk of depression actually increasing? *Journal of Clinical Epidemiology, 48* , 1109-1118.

Slutske, W. S., Heath, A. C., Dinwiddie, S. H., Madden, P. A., Bucholz, K. K., Dunne, M. P., Statham, D. J., et al. (1997). Modeling genetic and environmental influences in the etiology of conduct disorder: a study of 2,682 adult twin pairs. *Journal of .Abnormal.Psychology, 106,* 266-279.

Wang, X. L., Sim, A. H. S., Badenhop, R. F., McCredie, R. M., Wilcken, D. E. L. (1996). A smoking-dependent risk of coronary artery disease associated with a polymorphism of the endothelial nitric oxide synthase gene. *Nature Medicine , 2,* 41-45.

Chapter 13

TWINNING, FERTILITY, AND THE OPPORTUNITY FOR NATURAL SELECTION IN A MICRONESIAN ISLAND POPULATION

Jane Underwood

Key words: twinning, Darwinian fitness, fertility, selective advantage, Chamorro

Abstract: Consistent with recent findings (e.g., Lummaa et al., 1998; Madrigal, 1995), the present study also finds enhanced Darwinian fitness of Chamorro mothers of twins (TBMs) in pre-World War II Gaum. Examination of completed reproductive histories of singleton birth mothers (SBMs) and of TBMs indicates that the latter produced on average more children, especially sons, over a longer reproductive lifespan and attained higher age at death than SBMs. Differential mortality of second-generation offspring lowered, but did not erase, this selective advantage.

1.   INTRODUCTION

Any selective advantage which might expectably accrue to mothers producing more than one child at each birth event (multiple birth mothers, MBMs) compared to mothers producing only singleton offspring (SBMs) has generally not been detected in human populations thus far examined. Explanations for failure to detect this predictable outcome in non-contracepting groups have pointed toward differential mortality of these mothers, extended succeeding birth intervals or cessation of subsequent reproduction by mothers following multiple birth events, or lower rates of survivorship to reproductive age among these offspring. One recent study (Madrigal, 1995) has identified enhanced Darwinian fitness among MBMs, based on completed reproductive histories of 149 Costa Rican mothers. Since her investigations were conducted among a non-contracepting group lacking marked socioeconomic differentiation and with essentially equal access to extant medical care, investigations in comparable populations may elucidate whether her findings represent a more widespread phenomenon. More recently, Lummaa et al. (1998) have presented a brief summary of results from studies of twinning in several pre-industrial era Finnish populations indicating enhanced reproductive success of MBMs on the archipelago of Åland and Åboland.

The present study is based on data drawn from a linked family record register covering 38,934 native Chamorros, including 27,998 born on Guam between 1900 and December 1941, the date when vital record registrations were disrupted by Japanese military occupation. Among this latter number, 191 pairs of twins, or 1.36 per cent of all these births, were recorded. In addition to these twin pairs, one set of quadruplets and four sets of triplets were also reported, but are not included as part of this study. Through the period of 1900-1941 the island of Guam and its population were under the administrative authority of the U.S. Navy.

Of the total 191 Chamorro twin pairs, 42 were unlike-sexed, thus recognizably dizygotic twins. Applying Weinberg's differential method, adjusted to the secondary sex ratio recorded for all births (106.6), it is estimated that a total of 82 dizygotic twin pairs (for whom some degree of genetic causality has long been posited) were present. For these twin pairs, mean maternal age at birth was around 31 years, mean total number of births to twin birth mothers (TBMs) was 8.49, with an average of 5.71 surviving to age 15 years or the end of the reporting period, and mean birth order at first multiple birth event 4.72. Comparison of mean birth intervals between offspring born to all high fertility (10+ livebirths) mothers with completed reproductive histories, i.e., born from 1880-1889, reveals a difference of more than three months separating births to SBMs in contrast to livebirths to TBMs and a significantly longer reproductive lifespan among these TBMs. Calculation of fitness and selection coefficients for all Chamorro TBMs and

SBMs with completed reproductive histories indicates a wider range of selective advantage to mothers of twins than has previously been recognized.

The continuing appearance of multiple birth events, most commonly twinning, at varying rates in a wide range of human populations poses a dilemma in terms of evolutionary theory. While biparous births have the potential to increase the completed reproductive performance (hence, potentially, fitness) of twin-birth mothers (TBMs) in comparison to that of women producing only one child at each birth (singleton birth mothers, SBMs), increased fetal, offspring and maternal mortality risks (outlined in Bulmer 1970) have often been argued to outweigh this latent enhanced fitness of TBMs. Even estimates of twinning rates are complicated by such issues as secular changes in rates (Vogel and Motulsky, 1989) or the "vanishing twin syndrome," the early loss of a twin conceptus leading to a singleton birth of the survivor, a phenomenon which may be involved in up to 15 per cent of all twin conceptions (Bryan, 1994). Further, the rate of dizygotic (and, most recently, monozygotic) twinning appears to vary in populations studied to date, further contributing to difficulties in rate comparisons. Of even greater concern to investigations into the fitness of TBMs is the need for data covering not only the total reproductive performance of affected mothers, but also of their surviving offspring. Finally, modern twinning studies in many populations are increasingly complicated on the one hand by the dissemination and use of reliable contraceptives and, on the other, by assisted fertility techniques that artificially induce polyovulation.

Particularly in light of the latter considerations, studies of the completed fertility performance of TBMs in "natural fertility" populations assume critical importance in addressing some of these concerns. Studies in historical demography have tended to derive mainly from European and American sources, many with extensive parish and census records available for the purpose. The present study, based on a linked family record register for the pre-World War II native Chamorro population of the western Pacific island of Guam, is intended to expand the existing twinning data base, to establish a baseline for the continuing study of changing patterns of twinning in this group, and to evaluate the differential fecundability hypothesis of twinning as a plausible mechanism for maintaining any selective fitness of TBMs.

2. METHODS

Materials used in this study derive from a linked family record register constructed from the Spanish census of 1897, ethnodemographic information, and from vital registrations which began to be recorded shortly after the U.S. Navy's capture of the island in 1898 and, since 1950, under civilian government direction. For the present purposes, only those births recorded between January 1900 and December 1941 are analyzed due to a hiatus in record registrations during the years of Japanese military occupation (1941-1944). Fortunately, this prewar period was marked by very few cases of permanent emigration (recorded in the Annual Reports of the Governor of Guam) in marked contrast to high levels of out-migration in recent decades, thus ensuring that virtually all births to Chamorro mothers during the former period are included in these data.

While all 27,998 births reported between 1900-1941 were examined to determine overall incidence of multiple births, maternal age and parity at time of multiple birth events, estimates of dizygotic and monozygotic twin birth rates, etc., a smaller subset was more intensively reviewed to obtain measures of multi-generational fitness. This smaller sample derives from the verified reproductive histories of women born between 1880-89, all with completed reproductive performance by 1941. Mean number of everborn children, maternal age at first and last birth, deaths among offspring by 1941, and number of children born by 1941 to these offspring are compared between SBMs and TBMs included in this data subset.

3.   RESULTS

As indicated in Table 1, the crude twinning rate for these 27,998 births is 6.82 per thousand. Of the total 191 twin births, 42 involved unlike-sex pairs which, following Weinberg's differential method[1], leads to an estimated rate of 3.82 (s.e.=0.49) monozygotic (MZ) and 3.00 (s.e.=0.46) dizygotic (DZ) twin pairs among all these twins. Formed from the early splitting of a single conceptus and the subsequent development of two embryos, MZ "identical" twins have been estimated to occur at about the same rate, 3.0 - 4.0 per thousand, in all studied populations, at least until very recently (Westergaard et al., 1997). In contrast, DZ twinning, the result of two separate inseminations of two ova by two spermatazoa at about the same time, is considered to have a heritable component and DZ twinning rates have been shown to vary by ethnic group and to be subject to environmental influences (Tong et al., 1997).

Following Morton's pioneering studies in Hawai'i (1967), twinning rates for a number of Pacific islanders, both in situ (summarized in Pollard, 1986) and living abroad (Pollard, 1995), have been published. Most of these studies appear to be based on analyses of census, hospital and vital registration sources and to rely on ethnic self-identification, with the

Table 1. Chamorro female fertility and secondary sex ratio (SSR) for births occurring between 1900-1941

Total number of births	27,998	
Multiple births - 191 twin pairs 4 triplet sets 1 quadruplet set		
Total multiple births	398	
Total singleton births	27,600	SSR=106.6
Number of mothers of twins (TBMs)	185 (N=370)	
Number of twin births	191 (N=382)	SSR=109.9
Total number of everborn to TBMs	1571	
Recurrence rate =	3.14	
Twinning rate =	6.82 per 1000 births	
Mean parity at first twin birth	4.72	
Mean maternal age at first twin birth	30.66	
Mean number of all children born to TBMs	8.49	

exception of one study in New Britain (Scragg and Walsh, 1970). The results reveal twinning rates (6.5 to 31.2) mostly higher than those characteristically very low levels found in many Asian, particularly Japanese, groups, ranging from 5.83 to 10.69 (Pollard, 1995, p. 927). A single report for Micronesians, drawn from United Nations reports, records a twinning rate of 9.3. Thus, the twinning rate identified in the present study falls at the lower end of the Pacific islander groups, most closely resembling the 6.5 rate calculated for American Samoa (n=14). Given the general constancy of the MZ twinning rate among populations, the ratio of DZ/MZ twinning can be more revealing of comparative similarities and differences. That ratio, 0.79 for the Chamorro data, far more closely aligns with the Asian and Malay materials reported in Bulmer (1970), which range from 0.40 to 1.40 and contrasts sharply with values calculated from the same source for Europeans (1.73 - 2.82).

In examining the completed reproductive histories (summarized in Table 2) of Chamorro women born between 1880-89, all of whom had attained the age of at least 52 years by 1941, the higher twinning rate (10.3) of this sample doubtless reflects the as yet incomplete reproductive performance of younger mothers who contributed far more singleton births to the total 27,998 born before 1941. For this smaller sample, however, the mean number of births to SBMs, 6.55, as of surviving offspring, 4.46, clearly indicates the foundation for the rapid population growth that characterized the entire population in the post-war years. Yet this high level of SBM completed fertility pales in comparison with the completed reproductive performance of TBMs of the same maternal birth cohort (shown in Table 3) whose mean number of everborn offspring, 8.97, despite greater mortality, survived at a rate ($\overline{X}$ = 5.34) sufficient to exceed that of the offspring of SBMs. However, the surviving adult offspring of TBMs had only produced on average a slightly larger number of offspring ($\overline{X}$ = 7.59) than had the children of SBMs ($\overline{X}$ = 7.50) by 1941.

Table 2. Chamorro female fertility and secondary sex ratio of all offspring to SBMs born between 1880-1889.

Total number of females	1,068	
Total number of females at risk	779	
Total number of infertile females at risk	56	
Total number of fertile females	723	
Total of singleton-only-birth mothers (SBMs)	691	
(Total of twin-birth mothers, TBM=32)		
Total number of births to SBMs	4,528	SSR= 104.9
Mean number of all births	6.55	
Mean maternal age at first birth	23.79	
Mean maternal age at last birth*	37.13	
Mean closed birth interval	31.6 months	
Mean maternal age at death	39.4 years	
Number of surviving offspring		
to age > 15/1941	3,084	
Mean number of surviving offspring	4.46	
Number of grandchildren born before 1942	4,740	
Mean number of grandchildren	7.50	

---

*excludes women producing only 1 child

Given the identical mean age at first birth (23.8 years) for both SBMs and TBMs, it appears that the higher completed fertility of the latter women was in large measure a product of their longer reproductive lifespans, although fertility in the later years was ameliorated by the prolonged mean time lapse of 3.53 years between the penultimate and final births, presumably due to declining fecundity in the perimenopausal period. It would be a mistake to wholly ignore the role of overall closed birth intervals which, statistically, seem to favor SBMs ($\overline{X}$ = 31.6 months) over TBMs($\overline{X}$ = 34.0 months). The latter value is grossly inflated by the inclusion of a single anomalous case[2] which, when removed, reduces the mean birth interval for TBMs to 28 months, although the difference by either calculation is not statistically significant. Further, it is unlikely that higher pre-adult and, particularly, infant mortality (40 per cent) among twins than among their singleton siblings (32 per cent), by removing ovulatory suppression effects of lactation, played an important role in the fertility performance of TBMs since supplementary infant feeding began by at least three months postpartum and weaning by six months (Thompson, 1942).

Table 3. Chamorro female fertility and secondary sex ratio of all offspring to TBMs born between 1880-1889

Total number of TBMs	32	
Total number of births to TBMs	287	SSR=120.8
Mean number of all births	8.97	
Mean maternal age at first birth	23.75	
Mean maternal age at last birth	39.38	
Mean closed birth interval	34.0 months	
Mean parity of first twin births	4.84	
Mean maternal age at death	44.6 years	
Number of surviving offspring		
to age>15/1941	171	
Mean number of surviving offspring	5.34	
Number of grandchildren born before 1942	243	
Mean number of grandchildren	7.59	

Maternal mortality can in no way be shown in these data to have had an adverse affect on the completed fertility of Chamorro women. To thecontrary, the mean age at death by 1942 among TBMs (44.64 years) markedly exceeded that of SBMs (39.42 years). This finding contrasts sharply with results reported from older studies and with more recent historical investigations (Gabler and Voland, 1994; Haukioja et al.,1989; inter alia). Moreover, in the Chamorro TBM sample, only one woman died within one year of delivery, with this death occurring in December 1918, at the height of the influenza epidemic on Guam. While it is tempting to ascribe this enhanced maternal survivorship to improved medical conditions, a reading of an ethnographic description of the conditions of childbirth as late as 1938-39 (Thompson, 1942) strongly countervene so facile an interpretation. In fact, U.S. Navy records (Annual Report of the Governor of Guam for FY 1940) reveal that only 15 per cent of the 799 recorded births to native mothers occurred in a hospital setting.

While other studies (reported in Bulmer, 1970; Eriksson, 1973; Gabler and Voland, 1994) have reported a lower secondary sex ratio (SSR) among twin pairs than for their siblings, the results of this study show an impressively elevated ratio of males to females among twin births to Chamorro mothers. For those twins with mothers born between 1880-89, pre-1941 differential mortality resulted in an even higher SSR among survivors: 15 of 38 males and 8 of 26 females, or SSR=187.5. The higher SSR of all offspring born to these high completed fertility level TBMs is consistent with the previously demonstrated J-shaped curve of SSR by parity and/or parental age (Underwood, 1995) and appears to be further supported by examining the SSRs of offspring born to 30 randomly selected high-fertility SBMs from the same maternal birth cohort. The latter women, all of whom were also born between 1880-89 and who had completed fertility of 10 or more everborn children, produced 182 male and 151 female offspring (SSR=120.5).

4.   DISCUSSION

As the accumulating body of twinning studies expands to include the results of investigations in an ever more diverse set of human populations living in a variety of conditions, recognition increases of the great diversity that exists in twinning patterns, past and present. While strictly speaking, only those traits, as dizygotic twinning, involving some degree of genetic influence are directly subject to natural selection, so monozygotic twinning is often disregarded in examining the relative selective fitness of twinning. But as mothers of twins, whether DZ or MZ, may contribute differentially to the gene pool of succeeding generations, the relative selective fitness of mothers of both kinds of twins is relevant to the study of evolution.

The present study expands the existing sparse literature on twinning rates in Pacific island populations and, because of the time depth possible through the use of historical materials, allows, inter alia, an estimate of the selective value of overall twinning in one specific native Pacific islander group. While the data set used in this study contains no information on zygosity, estimated values (3.82 MZ and 3.00 DZ) for each kind of twin pair, as well as overall (6.82), fall at the lower end of Pacific and of worldwide ranges. Missing also from this family record register are data relating to prenatal loss and to stillbirth rates, and it is doubtful that most stillbirths or abortions were reported unless, as happened rarely during these years, the mother was hospitalized at the time of the event. Unfortunately, too, it is not possible to examine here any differential features in time lapse between marriage and first birth for, as I have previously reported (Underwood, 1990), a majority of all first births likely represented the outcome of pre- or extra-marital unions.

Despite its distance and very different history, this population overall appears to have shared the "...presence of some Western medical care and the lack of Western contraceptive measures..." (Madriga,l 1995, p. 785) described for Costa Rican post-reproductive mothers studied in Limon. On Guam, by the late 1930s, all but about 400 Chamorros (Rogers, 1995) continued to follow the Roman Catholic faith introduced by some of the early Spanish visitors after Magellan's discovery of the island in 1521 and promulgated assiduously following the establishment of a permanent mission in 1668. Church teachings, as well as the non-availability of reliable contraceptives, plus positive traditional values surrounding the desirability of large families, all served to encourage high fertility. With the exception of perhaps a dozen large land-holding, relatively more well-to-do, intermarrying families, mainly of Spanish-mestizo descent (Thompson, 1942), economic differentiation among the vast majority of Chamorros was minimal. Even job opportunities provided by naval authorities involved such low wage scales for native employees that dramatic changes in economic status from this source were severely limited during the prewar years.

Among significant changes introduced under the aegis of the U. S. Navy administration was the establishment of a public school system and some improvement of medical and health conditions including inoculation and anti-helminthic programs among Guam's schoolchildren. Improvements in the island's infrastructure with public health effects (roads, water treatment, food storage facilities, sewage handling), while primarily addressing U. S. Navy needs, to some extent benefited those Chamorros living in and around areas of naval facilities and settlements. For those living in the more remote areas of the island, these changes were scarcely noticeable until after the end of World War II (an all-weather road around the island was not completed until the 1960's).

As among the women of Limon whose reproductive histories revealed a statistically significant difference between the mean number of births to

SBMs (7.0) and to TBMs (9.7), Chamorro women evinced a similar pattern of statistically significant difference in mean number of everborn offspring (SBMs=6.55, TBMs=8.97, t=3.58). Higher mortality levels among Chamorro pre-adult children, however, reduced this differential so that the mean number of surviving offspring of TBMs (5.34) exceeded that of SBMs (4.46) by slightly less than one child. Thus, in comparison to the Limon women (Table 4), the relative fitness of Chamorro mothers born between 1880 and 1889, measured only in terms of the mean number of surviving offspring, is lower (cf Madrigal, 1995, p. 782).

More accurately, of course, Darwinian fitness is measured in terms of change "from one generation to the next generation of *parents*" (Bodmer and Cavalli-Sforza 1976, p. 757; italics mine), necessitating data covering at least three generations which, given the lifespan of most investigators, requires historical records analysis. In one such study, Gabler and Voland (1994) analyzed a family reconstitution register for 18th and 19th century Krummhort (Germany) to develop an estimated number of liveborn grandchildren to 180 TBMs and 2, 476 MBMs. As in the Chamorro study, mothers of twins differed significantly from SBMs in age at last birth (TBMs=38.86 years, SBMs=37.38 years), in mean number of offspring (TBMs=6.61, SBMs=4.31), and, more to the issue of offspring reproductivity, in estimated number of grandchildren (TBMs=9.85, SBMs=8.68). Using the procedure applied above in Table 4, the multi-generational fitness(s) of Krummhorn SBMs was calculated as only 0.88 (8.68/9.85). A similar, but more marked trend appears in the Chamorro data, with the relative multigenerational fitness of Chamorro SBMs attaining a selective value of 0.98 (7.50/7.59). In the latter case, however, completed reproductive performance of offspring is not completely known, so this measure of fitness is at best a crude estimate.

In summary, Chamorro TBMs on average had longer reproductive lifespans, experienced lower rates of mortality, produced more children, particularly sons, and had slightly higher intergenerational fitness than did the mothers of singleton-only offspring. So far as I have been able to determine, this population may well have been unique in the higher secondary sex ratio of twins and their siblings, as also in the higher SSR of survivors among the progeny of TBMs (and, perhaps, in the higher birth order progeny of SBMs) than has hitherto been reported from other twinning studies. These features are, however, wholly consistent with previously described SSR and male fertility patterns (Underwood, 1993, 1994, 1995, 1997) among the native population of prewar Guam and may even represent a part of a "Micronesian pattern" of fertility (Brewis and Underwood, 1994; Underwood and Brewis, 1995).

Table 4. Fitness and Selective coefficients of SBMs and TBMs: Limon* (Costa Rica) and Guam (Mariana Islands)

| | Limon | | Guam | |
	TBMs	SBMs	TBMs	SBMs
Mean number of surviving offspring	7.6	5.7	5.34	4.46
Fitness (W)	1	0.75	1	0.84
Selective coefficient(s)	0	0.25	0	0.16

While the present study does not provide support for the existence of any absolute difference in fecundability, Chamorro TBMs retained higher levels of the capacity to conceive during the terminal years of their reproductive lifespans, presumably reflecting slower rates of change in their production of gonadotropins (Vogel and Motulsky, 1986). As the intergenerational productivity of these mothers and their offspring produced a slight (2 per cent) advantage, it is possible that their contribution to succeeding postwar generations will be further enhanced under improved health and medical conditions now prevailing. Given the observed high SSR and increased survivorship of higher birth order offspring (see Figure 1), a considerable degree of constricted transmission of mtDNA must already have taken place. Studies covering the intervening 50-year period since 1941 are now underway to examine, inter alia, the effects of access and acceptability of modern contraceptive methods, as well as declining mortality levels, on twinning rates and fertility performance in this unique island population.

Figure 1.

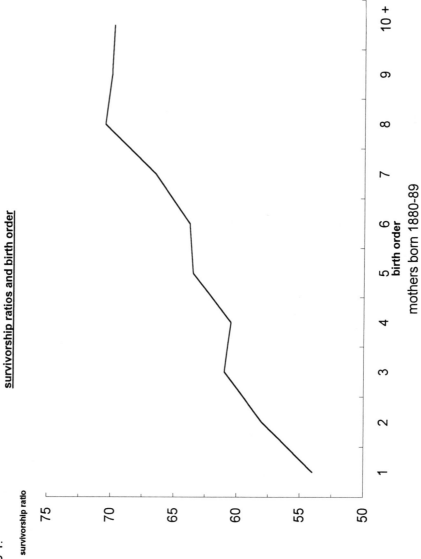

survivorship ratios and birth order

ACKNOWLEDGEMENTS

I am deeply grateful to the people of Guam whose enthusiastic cooperation and assistance have made this study possible and to Dr. Loren Madrigal and Dr. Stephen Molnar for their encouragement and provision of materials.

REFERENCES

Annual Reports. (1901-1940). *Annual Report of the Governor of Guam.* Washington, D.C.: U. S. Department of Navy.

Bodmer, W. F. & L. L. Cavalli-Sforza. (1976). *Genetics, Evolution, and Man.* San Francisco: W. H. Freeman & Co.

Brewis, A. 1993. Sex ratios at birth in a Micronesian atoll population. *Social Biology, 40,* 207-214.

Brewis, A. and J. Underwood. (1994). Sex ratios at birth in Micronesia: Reply to James. *Social Biology, 41,*280-282.

Bryan, E. (1994). Trends in twinning rates. *Lancet, 343,* 1151-2.

Bulmer, M. G. (1970). *The Biology of Twinning in Man.* New York: Clarendon Press.

Eriksson, A. W. (1973). *Human Twinning In and Around the Aland Islands.* Helsinki: Societas Scientiarum Fennica.

Gabler, S. & E. Voland. (1994). Fitness of twinning. *Human Biology, 66,* 699-713.

Haukioja, E., R. Lemmetyinen & M. Pikkola. 1989. Why are twins so rare in *Homo sapiens? American Naturalist, 133,* 572-577.

Lummaa, V., E. Haukioja, R. Lemmetyinen & M. Pikkola, (6 August 1998). Natural selection on human twinning: scientific correspondence. *Nature, 394,* 533-534.

Madrigal, L. (1995). Differential fertility of mothers of twins and mothers of singletons: study in Limon, Costa Rica. *Human Biology, 67,* 779-787.

Morton, N. E., C. S. Chung & M.-P. M. (1967). Genetics of interracial crosses in Hawaii. *Monographs in Human Genetics,* Vol. 3. New York: S. Karger.

Pollard, R. (1986). Twinning rates in the South Pacific, in I. S. Francis, B. F. J. Manley & F. C. Lam (Eds) *Proceedings of the Pacific Statistical Congress .* North-Holland: Elsevier Science Publishers.

Pollard, R. (1995). Ethnic comparisons of twinning rates in California. *Human Biology , 67,* 921-931.

Rogers, R. F. (1995). *Destiny's Landfall: a History of Guam.* Honolulu: University of Hawaii Press.

Scragg, R. F. R. & R. J. Walsh. (1970). A high incidence of multiple births in one area of New Britain. *Human Biology, 42,* 442-449.

Thompson, L. (1942). *Guam and its People.* San Francisco: American Council, Institute of Pacific Relations, Inc.

Tong, S., D. Caddy & R. V. Short. (March 22, 1997). Reports on the use of the dizygotic to monozygotic twinning ratio as a measure of fertility. *Lancet, 349,* 843-6.

Vogel, F. and A. G. Motulsky. (1986). *Human Genetics: Problems and Approaches.* Berlin: Springer-Verlag.

Underwood, J. H. (1990). Marital infertility and primary sterility in a native Pacific island population. *Micronesica , 23 ,*119-130.

Underwood, J. (1993). Secondary sex ratios in Micronesian populations. *Social Biology, 40,* 200-206.

Underwood, J. (1994). Seasonality of the secondary sex ratio in Micronesian island populations. *Journal of the Polynesian Society, 103,* 43-52.

Underwood, J. (1995). Sex ratios of livebirths in Micronesia. *American Journal of Human Biology, 7,* 241-249.

Underwood, J. (1997).   They also serve: Chamorro male fertility in the pre World War II period. *American Journal of Human Biology,* in press.

Underwood, J. and A. Brewis. (1995). A response to James "Coital rates, sex selective infanticide and sex ratios at birth." *Social Biology, 43,* 134-136.

Westergaard, T., J. Wohlfahrt, P. Aaby & M. Melbye. (March 15, 1997). Population based study of rates of multiple pregnancies in Denmark, 1980-94. *British Medical Journal, 7083 ,*775-779.

Notes

[1] Not corrected here for sex ratio in order to allow comparisons with other studies.

[2] In this case, involving a woman never in a consensual or formally married union, 18 years separated her first and second (final) parturition.

Chapter 14

ASSOCIATIONS BETWEEN THE ENDOTHELIAL NITRIC OXIDE SYNTHASE GENE (NOS3), REPRODUCTIVE BEHAVIORS, AND TWINNING

James MacMurray, Bruce Kovacs, Matt McGue, James P. Johnson, Hezekiah Blake, and David E. Comings

Key words: NOS3 gene, fertility decline, preeclampsia, twinning, menarche

Abstract: Fertility decline in industrialized nations, fuelled by increased educational opportunity and delayed marriage/childbearing, is a driving force in reshaping contemporary population structure. These same forces have contributed to an increase in twinning, due to fertility drugs, but a possible decline in natural dizygotic twinning. Twinning has been linked to elevations in testosterone and age-related increases in follicle-stimulating hormone. Using data from 473 females, we examined the role of the regulatory gene for nitric oxide synthase (NOS3) in fertility-related events and in the vulnerability to preeclampsia, or pregnancy-induced hypertension (PIH). We found that the 2-allele of a biallelic marker for this gene was associated with increased risk of PIH, with testosterone, with LH/FSH ratio, and with earlier age at menarche. Subsequent studies of the mothers of twins revealed that the NOS3 2-allele was associated with earlier maternal age at the birth of the twins, less education, and increased monozygotic twinning rate among females with birth ages <30. Additionally, 2-allele carriers evaluated by the multiphasic personality questionnaire had elevated scores for aggression and decreased scores for achievement orientation. These data suggest that, other events equal, the NOS3 2-allele carriers would outproduce NOS3 1-allele carriers, eventually threatening the viability of the 1-allele. However, prior to recent advances in medicine, preeclampsia, linked to the 2-allele, is thought to have had a roughly 40% rate of maternal/fetal mortality, balancing the selective advantages otherwise conferred upon the 2-allele.

1. INTRODUCTION

The Twentieth Century has witnessed a near-global transition from rapidly accelerating fertility and the spectre of mass starvation, to a decline in fertility that now finds many industrialized nations at below-replacement levels of reproduction. Following on the fears of a "population explosion" during the 1950s and 1960s, fertility in the U.S. and United Kingdom have subsequently been at a level below that needed for natural replacement of their populations for more than twenty years (Armitage & Babb, 1966). In addition to the overall trend toward a gross reduction in the average number of offspring produced by the average couple, the fertility transition is also characterized by an increasing tendency to delay the initiation of childbearing. Thus, the percentage of women giving birth in their 40s doubled in the past decade, and it is estimated that the proportion of total births in developed countries by the 35- to 49-year age group will increase by 72 percent, from 5.9 percent in 1982 to 8.6 percent by the turn of the century (Hansen, 1986). Moreover, delayed childbearing tends to coincide with both a declining ability to conceive, on the one hand, and increased medical complications of pregnancy on the other. Thus, the 1995 National Survey of Family Growth found that the absolute number of fecundity impaired women in the U.S. grew by nearly 30% just during the period between 1988 and 1995, an increase largely accounted for by a more educated baby-boom cohort, many of whom had delayed childbearing until their less fecund reproductive years (Chandra & Stephen, 1998). Moreover, the proportion of women in England and Wales who were still childless at age 45 rose to 13% in 1994, and is expected to increase further in the future (Armitage & Babb, 1996).

The trend toward delayed marriage and childbearing, as well as the option of remaining childless, has in large part issued from the major increases during this century in the average number of attained years of education. As underscored in the chapter by Comings (2000), individuals who do not complete high school initiate childbearing at between 20.5 and 22.8 years of age, whereas college graduates initiate childbearing at a mean of 26.5 years. Thus, the availability of educational opportunities and the increasing tendency to maximize those opportunities is a major driving force in the reshaping of contemporary population structure.

Paralleling the fertility transition, during recent decades there have also been reported declines in the sex-ratio (Dodds & Armson, 1997). Also, while the overall rate of multiple gestations has markedly increased as a result of the use of fertility drugs (Imaizumi & Nonaka, 1997), the 'natural' dizygotic (DZ) twinning rate has been reported to be in decline (Hur et al, 1995). These features may be linked, since twin births tend to show a lower sex-ratio than singletons (Beigulman et al., 1995). However, because the rate of DZ twinning has been shown to increase with maternal age (Lambalk & Schoemaker, 1997), the tendency toward later childbearing should lead to an increase, rather than a decrease, in DZ twinning rate. This apparent contradiction may at least in part be explained by the fact that the twinning rate is highest at parity four (Sharma, 1997), which would mean that advanced maternal age of primiparous, but not multiparous, females should be associated with a lowering of the DZ twinning rate.

The present study of genetic aspects of reproductive behaviors and twinning emerged as a consequence of studies being undertaken in our laboratory bearing on the genetic basis of preeclampsia, or pregnancy-induced hypertension (PIH). Preliminary results, to be reviewed below, indicate that the gene encoding endothelial nitric oxide synthase (NOS3), in addition to being associated with PIH, exerts regulatory influence over several reproductive hormones as well as menarcheal age. The nature of these findings suggested to us that this gene might have a broader role in mediating reproductive behavior in general, and the phenomenon of twinning in particular.

1.1 Nitric Oxide Synthase

Nitric oxide synthase (NOS), the catalytic agent for synthesis of nitric oxide (NO), exists in three established isoforms: NOS1, a neuronal form; NOS2, an inducible form; and NOS3, an endothelial form (Forstermann et al., 1998). NO is a gas molecule which, in addition to its role in blood vessel function (and penile erection), macrophage function, and the mediation of reproductive events, is also considered a neurotransmitter found in high densities in emotion-regulating brain regions (Nelson et al., 1995). Targeted disruption of the neuronal isoform has been shown to produce both exaggerated aggression and hyper-sexualized behaviors (Demas et al., 1997), but only in male animals (Kriegsfeld et al., 1997). Conversely, the endothelial NOS isoform NOS3 has principally been studied in females, since variations in levels of that enzyme have been shown to influence critical dimensions of reproductive biology (Buhimschi et al., 1998). Collectively, these findings indicate that the NOS genes may have special relevance for human studies of reproductive function and aggression.

1.2 Preeclampsia Studies

Pregnancy-induced hypertension (PIH) is a systemic endothelial disease that produces an imbalance between vasoconstrictive and vasodilating circulating hormones (Fournier et al, 1995), complicates roughly 6-8% of all gestations, and is the leading cause of both maternal and fetal death (Buschimi et al., 1998). In the past several years there has been a growing interest in the possible role of nitric oxide (NO) deficiency in the pathophysiology of PIH (Baylis et al., 1998). NO is a potent vasodilator released by endothelial cells and synthesized by the catalytic action of the endothelial constitutive NO synthase (eNOS) (Nasiell et al., 1998). Rats infused with L-NAME, an NOS inhibitor, have been used as an animal model of PIH, leading to the view that PIH may represent a state of NO deficiency (Buhimschi et al., 1998). A disturbed NO/endothelin-1 equilibrium in cultured human placental endotheliocytes in PIH has been reported (Leszcynska-Gorzelak et al., 1998), supporting the hypothesis that PIH entails insufficient NO release in response to a stimulus. Moreover, the potential role of the gene encoding eNOS (NOS3), at chromosome 7q36, was evaluated by linkage analysis among affected sisters in multiplex families (Arngrimsson et al., 1997), revealing a distortion from expected allele-sharing with a LOD score of 3.36 (p=.005). To date, however, molecular genetic studies of the NOS3 gene in PIH have not been undertaken, which prompted us to examine the NOS3 marker in the following series of studies.

1.3   The NOS3 Gene, Reproductive Hormones, and PIH

In our initial study we examined the NOS3 single base pair polymorphism in samples of adult female Hispanics obtained from the Women's Hospital at the University of Southern California. Sample one was comprised of 117 females unscreened for PIH, age range 21-43, mean age 24.6. Within this sample, age at onset of menarche was determined for 112 subjects, and for 78 of the subjects we obtained basal concentrations of free testosterone (T), progesterone (P4), leutenizing hormone (LH), and follicle stimulating hormone (FSH), measured at mid-luteal phase. In addition, the ratio of LH to FSH, which is considered a measure of fecundability, was evaluated. Sample two was comprised of 28 subjects with a history of pregnancy-induced hypertension, and sample three was comprised of 21 subjects with two or more previous pregnancies with no evidence of PIH. Blood samples were obtained by forearm venipuncture, and the samples were transported to the Department of Medical Genetics at the City of Hope National Medical Center for genetic analysis. The samples were centrifuged and the DNA pellet extracted using standard protocols. NOS3 genotyping was performed using the protocol described by Wang et al., 1996).

In Table 1, below, the distribution of genotypes are shown for the three groups examined. There was no difference between the unscreened sample and the non-PIH sample; whereas the PIH group had a significant excess of 2-allele carriers when compared either with the non-PIH group (Pearson, $p<.05$) or the unscreened sample ($p<.001$). This association is currently being reevaluated in larger samples of PIH and non-PIH subjects, but the preliminary findings suggest that NOS3 2-allele carriers, and particularly 22-homozygotes, are at increased risk of developing PIH.

Table 1: NOS3 Genotype Distributions in Three Samples of Hispanic Females

	UNSCREENED	NON-PIH	PIH CASES
NOS3-11	92 (78%)	17 (81%)	13 (46%)
NOS3-12	22 (19%)	4 (19%)	11 (39%)
NOS3-22	3 (3%)	0 (0%)	4 (14%)

In our evaluation of the hormone profiles of the 78 unscreened Hispanic females, 2-allele carriers were significantly different from 11-homozygotes on three of the six comparisons made. These results are shown in Table 2. The hormone levels are measured in milligrams per deciliter.

Table 2: NOS3 Associations with Reproductive Features in 78 Hispanic Females

	TESTOSTERONE	LH/FSH RATIO	MENARCHEAL AGE
NOS3-11	65.3 (+/- 32.4)	1.18 (+/- .78)	13.30 (+/- 1.78)
NOS3-12	93.6 (+/- 56.1)	2.26 (+/- 1.80)	12.90 (+/- 1.86)
NOS3-22	46.0 (+/- 0)	2.00 (+/- 0)	10.00 (+/- 1.41)
	$p=.037$	$p=.048$	$p=.031$

These findings suggest that NOS3 2-allele carriers undergo earlier sexual development than 11-homozygote females. Other studies have shown that menarcheal age is positively and significantly related to age at first sexual intercourse, first pregnancy, and first live birth (Scholl et al., 1989). The elevated concentrations of testosterone observed in association with the 2-allele is of interest for several reasons. First, elevated testosterone levels in females is associated with increased disposition to twinning (Thomas et al., 1998). Additionally, female testosterone levels have been found to be significantly correlated with measures of female dominance (Cashdan, 1985) as well as with aggression and number of sex partners (van Goozen et al., 1995); and both maternal testosterone (Abeliovich et al., 1984) and female dominance (Grant, 1996) have been found to eventuate in sex-ratio distortion in offspring, namely an excess of male births.

Our finding of an increased ratio of LH-to-FSH in 2-allele carrying women is of particular relevance to our interest in twinning. In general, monozygotic (MZ) twinning occurs at a rate of around 3-4 per thousand maternities in all populations (Imaizumi & Nonaka, 1997), and until recently the MZ twinning rate was thought to be largely invariant. On the other hand, the disposition to DZ twinning is highly heritable, and varies widely with race and maternal age (Tong et al., 1997). During pregnancy, women carrying twins have been reported to have a 50% higher mean concentration of testosterone and a 49% higher mean concentration of FSH than women carrying singletons (Thomas et al., 1998), and an elevated FSH is considered to be the principal underlying mechanism in DZ twinning in particular (Lambalk & Schoemaker, 1997; Martin et al., 1984). In turn, FSH concentrations undergo a normal increase in average levels during the reproductive lifespan, which is thought to increase the probability of DZ twinning in older females (Reame et al., 1998). Lastly, of course, higher rates of multiple gestations occur in females being treated with ovulation-inducing hormones, which increase FSH levels (Imaizumi & Nonaka, 1997), and which accounts for the overall rise in the DZ births to older females.

The higher concentrations of testosterone in NOS3 2-allele carrying females suggests that they might have a heightened tendency toward multiple gestations, whereas the lower FSH-LH ratio suggests they would be less likely to produce DZ twins. Therefore, we hypothesized that NOS3 2-allele carrying females would have an increased rate of twinning in general, and MZ twinning in particular. Also, on the basis of their increased testosterone levels we hypothesized that these women would demonstrate increased aggression, and would produce an excess of male offspring. In addition, on the basis of their earlier age at menarche, we hypothesized that they would begin childbearing at an earlier age. Moreover, in view of the positive correlation between age at initiation of childbearing and years of education, we hypothesized that 2-allele carrying females would, in association with earlier childbearing, have truncated educational attainment and reduced achievement orientation, compared with 11-homozygote females.

1.4 Studies of Twinning and Behavior

In order to test the hypotheses outlined above we examined a total of 549 non-Hispanic Caucasian parents of same-sex twins from the Minnesota Twin Family Study (MTFS). Although the principal focus of this study was the possible role of the NOS3 gene in female reproductive behaviors, we also sought to explore its relevance in males. The sample was comprised of 307 mothers and 242 fathers. 324 (59.0%) of the twins born to these subjects were MZ and 225 (41.0%) were DZ; 418 (76.1%) were male twin sets and 131 (23.9%) were female twin sets. The age range of mothers was 31 to 57, with a mean of 41.2; their age at the birth of the twins had a range of 17 to 40, with a mean of 27.1. The age range of the fathers was 32 to 63, with a mean of 44.3; their age at the birth of the twins had a range of 20 to 46, with a mean of 30.1. The number of children born to the parents prior to the birth of the twins was unknown. In order to control for possible effects of assortative mating, 68% of the parents were unrelated (unmarried), while the 32% who were married couples allowed us to directly examine the possibility of assortative mating at this locus. 74% of the subjects were administered the Multiphasic Personality Questionnaire (MPQ), which contains aggression and achievement orientation scales. We also included in our analysis the subjects' age at the birth of the twins, educational level, twin zygosity and twin-sex. Blood samples obtained from the subjects were shipped to the City of Hope, and genetic studies were performed using the protocols described earlier.

2. RESULTS

The distribution of NOS3 alleles in the MTFS sample was similar to that observed in the Hispanic females described above. The genotypes for males and females are shown in Table 3. There was no evidence of assortative mating at this genetic locus, and the NOS3 genotypes and allelic frequencies for males and females are nearly identical. Moreover, an examination of the NOS3 gene in fathers of MZ and DZ twins revealed that they too were identical. However, a similar examination of mothers of MZ

Table 3: Distribution of NOS3 genotypes in samples of parents of twins.

	NOS3-11	NOS3-12	NOS3-22
MALES	173 (71%)	62 (26%)	7 (3%)
FEMALES	213 (69%)	78 (25%)	16 (5%)

and DZ twins found that MZ mothers had an excess of 2-alleles when compared either with DZ mothers (p<.02) or with fathers (p<.05). No NOS3 differences were found for zygosity or any other trait in male subjects. Therefore, all further reviews are limited to the findings in mothers. The association between maternal NOS3 genotype and zygosity of twins is shown in Table 4.

Table 4: Zygosity: Maternal NOS3

	NOS3-11	NOS3-12	NOS3-22
MZ TWINS	117 (65%)	51 (28%)	13 (7%)
DZ TWINS	96 (76%)	27 (21%)	3 (2%)

PEARSON r, p=.048, LINEAR BY LINEAR TREND, p=.015

In our next analyses we examined the relationship of the NOS3 gene to twin-sex and maternal age at the birth of the twins. The gene marker was found to have no association with the gender of the twins but, as predicted, female carriers of the 2-allele gave birth to their twins at an earlier age than 11-homozygotes (p<.025). These data are shown in Table 5.

Table 5: Birth Age: Maternal NOS3

	NOS3-11	NOS3-12	NOS3-22
AGE 24<	47 (57%)	29 (35%)	6 (7%)
AGE 25-29	83 (72%)	27 (24%)	5 (4%)
AGE 30-34	57 (75%)	15 (20%)	4 (5%)
AGE 35>	26 (77%)	7 (20%)	1 (3%)

Because the twinning rate increases with age, we next examined mothers of MZ and DZ twins separately for interaction between the NOS3 gene and maternal birth age, and for association with the ratio of MZ-to-DZ twin offspring. Among those women who gave birth to their twins at age 30 or older, there was no association between the NOS3 alleles and zygosity of the twins. However, among those who gave birth prior to age 30, the percentage of MZ twins, as a proportion of MZ and DZ twins born to mothers, was: 11- 54%, 12-66%, 22 100% (p<.005). These data are shown in Figure 1.

```
                                     NOS3 GENE
                    Count
                    Row Pct      11        12        22
                    Col Pct                                    Row
                                                               Total
  ZYGOSITY        ─────────────────────────────────────────
                                70        37        11         118
                    MZ          59.3      31.4      9.3        59.9
                                53.8      66.1      100.0

                                60        19                   79
                    DZ          75.9      24.1                 40.1
                                46.2      33.9

                    Column      130       56        11         197
                    Total       66.0      28.4      5.6        100.0
```

Chi-Square	Value	DF	Significance
Pearson	10.23527	2	.00599
Likelihood Ratio	14.13688	2	.00085
Mantel-Haenszel test for linear association	9.03585	1	.00265

Figure 1: Percent Monozygotic Twins by NOS3 Genotype

These data suggest that the 2-allele carrying females have their twins at an earlier age, and those early pregnancies produce an excess of MZ twins; whereas 11-homozygote females have their twins at later ages, but appear to have an expected MZ/DZ twinning ratio. That is, the only imbalance in twinning ratio observed appears to be an excess of MZ twins born to young 2-allele mothers.

2.1 NOS3 and Personality Traits

Given the earlier maternal age at twin birth observed above for 2-allele carriers, it might be expected that the choice to have children early would interfere with education. Consistent with such a model contrasting early childbearing with completion of educational objectives, we found that in an examination of all mothers, 2-allele carrying females have less education than 11-homozygote females (p<.025). However, this difference was again entirely accounted for by the mothers of MZ twins, where the association was more robust (p=.0055). These data are shown in Table 6.

Table 6. Association between the NOS3 Gene and Educational Level in Mothers of MZ Twins.

	NOS3-11	NOS3-12	NOS3-22
HS OR LESS	67 (59%)	37 (32%)	10 (9%)
ASSOC. DEGREE	8 (53%)	5 (33%)	2 (13%)
BA/BS	35 (78%)	9 (20%)	1 (2%)
PROFESSIONAL	7 (100%)	0 (0%)	0 (0%)

In the next analysis mean MPQ scale scores for Aggression and Achievement were analyzed separately for mothers of MZ and DZ twins. Again, NOS3 associations with behavioral features were limited to the mothers of MZ twins. As shown in Table 7, females who carry the 2-allele had significantly elevated mean Aggression scores and significantly lower mean Achievement scores, relative to 11-homozygotes.

Table 7: Aggression and Achievement by NOS3 Genotype

	AGGRESSION	ACHIEVEMENT
NOS3-11	26.3 (+/- 5.0)	50.1 (+/- 7.7)
NOS3-12	29.4 (+/- 5.2)	46.0 (+/- 6.4)
NOS3-22	31.8 (+/- 4.9)	45.2 (+/- 5.6)
	F:6.615, P=.002	F:4.44, P=.014

The significant dose-dependent differences in Aggression and Achievement scores, as well as educational level, linked to the number of 2-alleles present, supports our hypothesis that the allelic association with testosterone would also be expressed in increased aggression in 2-allele carriers. Since aggressive behaviors tend to decline with age, as does twinning, we examined the interaction between age and aggression scores in mothers of MZ twins, by NOS3 genotype. As shown in Figure 2, the mean Aggression scores do in fact decline with age for all three NOS3 genotypes. As a result of the age-related convergence of scores, the NOS3 effect is entirely restricted to the younger female subjects.

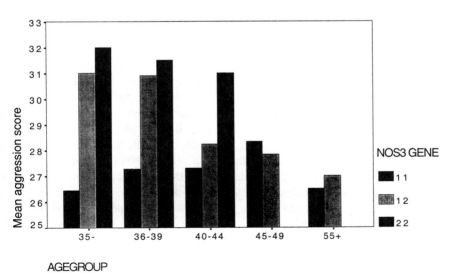

Figure 2: Mean MPQ Aggression Score by Age by NOS3 Genotype in
Mothers Who Have Their Twins at Less than 30 Years of Age

3. SUMMARY

The data reviewed here suggest that the endothelial nitric oxide synthase gene exerts important regulatory influence over female reproductive events. The association between this gene and both age at menarche and reproductive hormone profiles in women is probably the critical intermediary step to its linkage to the more complex human traits and behaviors observed in these studies. Collectively, these findings suggest that females who carry the 2-allele of the NOS3 polymorphism have an earlier age at menarche, may initiate childbearing at an earlier age, are more aggressive and less achievement-oriented, are less well educated, and have a larger proportion of MZ twins during their early reproductive careers than those females who are NOS3 11-homozygotes. Conversely, it would appear that it is the 11-homozygote female who exemplifies the demographic trend toward increased education and its corrolary, delayed childbearing.

Prior to the recent trend toward delayed childbearing, it would be expected that the earlier age at menarche and childbearing observed in NOS3 2-allele carrier females would give them a selective advantage with regard to total reproductive capacity. That is, all other things being equal, 2-allele carrying females ought to produce a greater total number of offspring than 11-homozygote females. If that were true, however, such a strong selective advantage would rapidly displace and eventually eliminate the NOS3 1-allele from most populations. Thus, from an evolutionary perspective, it might be expected that a countervailing, or balancing trait, such as that witnessed with the sickle cell trait and protection from malaria, would be involved at this locus in order to maintain the NOS3 1-allele in human populations. To that end, the apparent greater susceptibility of NOS3 2-allele females to preeclampsia would probably serve such a purpose. It is estimated that prior to the advent of medical treatment for preeclampsia, which occurs in 6 to 8% of the population, this disorder had maternal and fetal mortality rates of roughly 40% (Buschimi et al., 1998). Moreover, since by definition preeclampsia is a disorder that affects primigravidae, a greater proportion of NOS3 2-allele females would theoretically be removed from reproductive competition before having the opportunity to pass on this allele to any offspring. Through such a mechanism, we propose, the NOS3 1-allele has been maintained in human populations.

REFERENCES

Abeliovich, D., Leiberman, J. R., Teuerstein, I. & Levy, J. (1984). Prenatal sex diagnosis: Testosterone and FSH levels in mid-trimester amniotic fluids. *Prenatal Diagnosis, 4,* 347-353.

Armitage, B. & Babb, C. (1996). Population review: (4). Trends in fertility. *Population Trends, 84,* 7-13.

Baylis, C., Beinder, E., Suto, T., & August, P. (1998). Recent insights into the roles of nitric oxide and reninangiotensin in the pathophysiology of preeclamptic pregnancy. *Seminars in Nephrology, 18,* 208-230.

Beiguelman, B., Franchi-Pinto, C., Dal Colletto, G. M., & Krieger, H. (1995). Annual variation of sex ratio in twin births and in singletons in Brazil. *Acta Genet. Med. Gemello (Roma), 44,* 163-168.

Buhimschi, I., Yallampalli, C., Chwalisz, K., & Garfield, R. E. (1995). Pre-eclampsia-like conditions produced by nitric oxide inhibition: Effects of L-arginine, D-arginine and steroid hormones. *Human Reproduction, 10,* 2723-2730.

Cashdan, E. (1995). Hormones, sex, and status in women. *Hormones and Behavior, 29,* 354-366.

Chandra, A. & Stephen, E. H. (1998). Impaired fecundity in the United States: 1982-1995. *Family Planning Perspectives, 30,* 34-42.

Cloninger, C. R. (1978). The antisocial personality. *Hospital Practice, 13,* 97-106.

Comings, D. E. (2000). Age at first childbirth: A major selective factor for psychiatric genes in the twentieth century. In J. L. Rodgers, D. C. Rowe, & W. B. Miller (Eds.) *Genetic Influences on Human Fertility and Sexuality,* Mahweh, NJ: Kluwer.

Demas, G. E., Eliasson, M. J., Dawson, T. M., Dawson, V. L., Kriegsfeld, L. J., Nelson, R. J., & Snyder, S. H. (1997). Inhibition of neuronal nitric oxide synthase increases aggressive behavior in mice. *Molecular Medicine, 3,* 610-616.

Dodds, L. & Armson, B. A. (1997). Is Canada's sex ratio in decline? *Canadian Medical Association Journal, 157,* 46-48.

Forstermann, U., Boissel, J. P., & Kleindert, H. (1998). Expressional control of the 'consistutive' isoforms of nitric oxide synthase (NOS I and NOS III). *FASEB J, 12,* 773-790.

Fournier, A., Fievet, P., el Esper, I., el Esper, N., Vaillant, P., & Gondry, J. (1995). Hypertension and pregnancy. Diagnosis, physiopathology and treatment. *Schweiz. Med. Wochenschr., 125,* 2273-2298.

Grant, V. J. (1996). Sex determination and the maternal dominance hypothesis. *Human Reproduction, 11,* 2371-2375.

Hansen, J. P. (1986). Older maternal age and pregnancy outcome: A review of the literature. *Obstetrics and Gynecology Survey, 41,* 726-742.

Hur, Y. M., McGue, M., & Iacono, W. G. (1995). Unequal rate of monozygotic and like-sex dizygotic twin birth: Evidence from the Minnesota Twin Family Study. *Behavior Genetics, 25,* 337-340.

Imaizumi, Y. & Nonaka, K. (1997). The twinning rates by zygosity in Japan, 1975-1994. *Acta Genet. Med. Gemello (Roma), 46,* 9-22.

Kriegsfeld, L. J., Demas, G. E., Lee, S. E., Dawson, T. M., Dawson, V. L., & Nelson, R. J. (1999). Circadian locomotor analysis of male mice lacking the gene for neuronal nitric oxided synthase (nNOS-/-). *Journal of Biological Rhythms, 14,* 20-27.

Lambalk, C. B. & Schoemaker, J. (1997). Hypothetical risks of twinning in the natural menstrual cycle. *European Journal of Obstetrics, Gynecology, and Reproductive Biology, 75,* 1-4.

Leszczynskaz-Gorzelak, B., Oleszczuk, J., Kaminski, I., Szymula, D., Gorzelak, M., & Oleszczuk, J. L. (1998). Disturbed nitric oxide/endothelin-1 equilibrium in cultured human placental endotheliocytes in preeclampsia. *Gynecologic and Obstetric Investigations, 46,* 145-149.

Martin, N. G., Beaini, J. L., Olsen, M. E., Bhatnagar, A. S., & Macourt, D. (1984). Gonadotropin levels in mothers who have had two sets of DZ twins. *Acta Genet. Med. Gemello (Roma), 33,* 131-139.

Nasiell, J., Nisell, H., Blanck, A., Lunell, N. O., & Faxen, M. (1998). Placental expression of endothelial constitutive nitric oxide synthase mRNA in pregnancy complicated by preeclampsia. *Acta Obstetrics and Gynecology Scand., 77,* 492-496.

Nelson, R. J., Demas, G. E., Huang, P. L., Fishman, M. C., Dawson, V. L., Dawson, T. M., & Snyder, Sl H. (1995). Behavioral abnormalities in male mice lacking neuronal nitric oxide synthase. *Nature, 378,* 383-386.

Reame, N. E., Wyman, T. L., Phillips, D. J., de Kretser, D. M., & Padmanabhan, Y. (1998). Net increase in stimulatory input resulting from a decrease in inhibin B and an increase in activin A may contribute in part to the rise in follicular phase follicle-stimulating hormone of aging cycling women. *Journal of Clinical Endocrinology and Metabolism, 83,* 3302-3307.

Scholl, T. O., Hediger, M. L., Vasilenko, P., Ances, I. G., Smith, W., & Salmon, R. W. (1989). Effects of early maturation on fetal growth. *Annals of Human Biology, 16,* 335-345.

Sharma, K. (1997). The twinning rates and epidemiological characteristics of births in southeast Uttar Pradesh, India. *Acta Genet. Med. Gemello (Roma), 46,* 47-56.

Thomas, H. V., Murphy, M. F., Key, T. J., Fentiman, I. S., Allen, D. S., & Kinlen, L. J. (1998). Pregnancy and menstrual hormone levels in mothers of twins compared to mothers of singletons. *Annals of Human Biology, 25,* 69-75.

Tong, S., Caddy, D., & Short, R. V. (1997). Use of dizygotic to monozygotic twinning ratio as a measure of fertility. *Lancet, 349,* 843-845.

Van Goozen, S. H., Matthys, W., & Cohen-Kettenis, P., T., Thijssen, J. H., & van Engeland, H. (1998). Adrenal androgens and aggression in conduct disorder prepubertal boys and normal controls. *Biological Psychiatry, 43,* 156-158.

Wang, X. L., Sim, A. H. S., Badenhop, R. F., McCredie, R. M., & Wilcken, D. E. L. (1996). A smoking-dependent risk of coronary artery disease associated with a polymorphism of the endothelial nitric oxide synthase gene. *Nature Medicine, 2,* 41-45.

GENETIC INFLUENCES ON HUMAN FERTILITY AND SEXUALITY: COMMENTARY ON CHAPTERS 10-14

A. J. Figueredo

As an evolutionary psychologist, my first reaction to these chapters was to be quite favorably impressed by the way that behavioral genetics, as exemplified by these works, has been incorporating the ideas of evolutionary psychology and fully integrating them into a truly synthetic biological perspective. Regrettably, my second reaction was to be chagrined over the relative unwillingness or inability of many of my fellow evolutionary psychologists to integrate behavioral genetic perspectives into their own ideas. Both of these reactions stem from my personal view that these approaches do not really represent two independent fields of study, but are instead two aspects of a single Darwinian behavioral science that is only now emerging the way that the original Neo-Darwinian Synthesis did over fifty years ago. I think that this is closer to the way that both Charles Darwin and our other common ancestor, Sir Francis Galton, envisioned the biological psychology of the future. Or, as my friend, colleague, and former graduate mentor, Lewis Petrinovich, once put it jestingly in private conversation: "We can have both a mean *and* a variance!"

Although there is much on which to comment in these excellent chapters, I will restrict myself to aspects I am most conversant with, which are the evolutionary implications of these behavioral genetic findings. The first of these chapters, by Dawood and Bailey, deals with the genetics of human sexual orientation. Aside from impressing the reader with the tremendous difficulty of obtaining a genetically-informative sample that is also representative, and the consequent need for the extremely careful research which the authors are evidently engaged in, one is also left with the unmistakable impression that the etiology of homosexual orientation must be a *bona fide* gene-environment interaction of nontrivial magnitude and major reproductive consequences. Although the evidence at this point unequivocally indicates some degree of genetic involvement, the relatively low rate of concordance for homosexuality even among *MZ* twins, although statistically different from zero, suggests that there must be some additional nonshared environmental trigger. If this is so, then one is inevitably drawn to ask what this trigger might be.

In addition, one cannot help but try to connect that to the fundamental paradox underlying the heritability of homosexuality, which is how such an evidently fertility-suppressing trait can possibly survive the force of natural selection. We are not talking here about some novel pathology brought on by modern metropolitan or industrial environments; homosexual behavior has been documented virtually worldwide in human history for several millennia. Various alternative explanations have been offered, such as the possibility that homosexual behavior occurred historically within the context of a socially tolerated and non-fitness-reducing bisexuality, but none of them are ultimately very satisfying. However, the gene-environment interaction hypothesis might offer some potentially fruitful avenues for research. What are the consequences of having the "homosexual genes" without developing the gay phenotype? Could these genes have other and more beneficial fitness consequences if the presumed environmental trigger is *not* present to alter sexual orientation. The average fitness payoff of carrying those genes could still be positive. Many characteristics which are usually considered socially desirable have been identified as statistically overrepresented in gay men, such as higher artistic talent, higher intelligence, higher socioeconomic status, higher law-abidingness and social conscientiousness, *etc.* While these might represent indirect consequences of expressing the full-fledged gay phenotype, I would suggest that researchers might be well-advised to start looking at the relative accomplishments of the nonconcordant ("straight") twins. If they share the fitness benefits presumably conferred by these associated traits without expressing the altered sexual orientation, then the suspected "homosexual genes" might be maintained within the population by the stochastic benefits of their pleiotropic effects.

The second of this set of chapters, by Trumbetta and Gottesman, bases its analysis of the behavioral genetics of marital status upon a similar concept, which is that of *endophenotypes*. These endophenotypes are not identical with the specific behaviors targeted, but are instead the psychological phenotypes that are thought to ultimately underlie those behaviors. In the specific case of tracking marital status, these endophenotypes are mating strategies characterized by either (1) high parental investment, or (2) high mate diversification. These two reproductive strategies appear identical to those described by behavioral ecologists as biased towards the allocation of total reproductive effort to either (1) high parental effort, or (2) high mating effort (*cf.*, Rowe et al., 1997). The paradox here is not that these endophenotypes show high heritability, because individual differences in reproductive strategy have long been suspected to have a partially genetic basis (*cf.*, Gangestad & Simpson, 1990), but that this heritability appears to be declining over time. The argument is made that the recent liberalization of divorce laws in the United States should be increasing heritability by attenuating environmental constraints, yet the reported results of the *NAS-NRC* sample show a decrease in heritability over this period. Once again, methodological concerns are raised in this chapter, such as the current inadequacy of the data to distinguish between age and cohort effects, and much future research will be needed to resolve this paradox. The major virtue of this chapter, however, does not reside in the finality of its conclusions, but in the careful attention devoted to identifying the more fundamental latent traits that might be underlying observed patterns of behavior.

The third chapter in this set, by Comings, also takes up the issue of what causal factors that are not intuitively obvious might be influencing the manifestation of a targeted behavior. In this case, the focus is on age at first childbirth. The argument here is that an earlier age at first childbirth has been systematically favored by natural selection in the United States over much of the 20th Century. This selective advantage is a natural consequence of the faster generation time produced by earlier reproduction. Furthermore, this trait is shown to be associated with a variety of psychiatric problems, such as conduct disorder. Thus, the selective advantages conferred by earlier reproduction is held responsible for the observed increase in prevalence of these psychiatric disorders. The psychiatric disorders that have increased in prevalence include substance abuse, anxiety disorders, obsessive-compulsive disorders, phobias, schizophrenia, and childhood learning disabilities. The observed increases in prevalence are therefore not directly attributable to changes in social or environmental factors, but to a microevolutionary process of changes in gene frequency within the population over time.

On the other hand, a social cause is postulated for conferring the said selective advantage on the early-reproduction genes and is therefore *indirectly* held responsible. This social factor is the increasing number of individuals, especially women, availing themselves of higher education in the United States over much of the 20th Century, which is correlated with delayed fertility. This is where the analogy with the Trumbetta and Gottesman chapter falls apart. Although the "endophenotype" for early age at first childbirth is well-described in the Comings chapter, with all its socially undesirable correlates, the corresponding "endophenotype" for higher education and delayed fertility is explored in less detail. This weakens the causal argument.

For example, a person with all the psychiatric problems described above as correlates of early reproduction is a poor candidate for completing any protracted course of higher education. The same holds for early pregnancy itself, which is cited in this chapter as a major cause of dropping out of school. Which, then, is the cause and which the effect? Is higher education really delaying the fertility of certain groups, or are the genes rendering individuals either unwilling or unable to participate in higher education merely increasing? By default, those individuals availing themselves of higher education would perforce be those lacking the "early reproduction genes" that interfere with it. Thus, the observed effect of higher education on delayed fertility might be nothing more than that of a gene-environment correlation (attributable to self-selection) and not a true environmental effect at all. To strengthen the causal argument proposed in this chapter, one would have to show that the availability of higher education decreased the fertility of the specific genotypes participating in it relative to *their own* previous levels and not relative to that of some alternative reproductive strategy.

Higher education is not distributed randomly, but is differentially utilized by different genotypes. For example, the chapter cites higher *IQ* as a strong correlate of higher education, something which most readers of this volume will doubtless recognize as a primarily genetically-determined trait. Because higher education is a means of achieving higher social status and increased material resources, it is therefore not unreasonable to suppose that those most highly motivated to avail themselves of it are individuals genetically committed to a "high parental investment" reproductive strategy. One of the hallmarks of this strategy is delayed fertility rather than early and exhaustive reproduction. The correlation between delayed fertility and higher education may therefore be entirely genetic and not based upon any external social or environmental causes. Of course, one could claim that highly *K*-selected individuals are more susceptible to the baneful lure of higher education, which is probably accurate. Nevertheless, one would still have to show that higher education actually decreased *their own* already low fertility. It is already axiomatic that their fertility will be significantly lower than that of *r*-selected individuals (*cf.*, Rushton, 1985).

Furthermore, one would have to show that it was higher education itself that was responsible. For much of the 20th Century, there has also been an increasing participation of women in the labor market. Over recent decades, this labor market has become increasingly competitive in terms of the skills required to perform high-paying jobs (*cf.*, Herrnstein & Murray, 1994). Increased competition over resources is one of the factors that classically favors *K*-selected reproductive strategies. Under such ecological conditions, the selective advantages conferred by early reproduction would be entirely negated in the absence of a welfare state that subsidized the reproduction of what would economically be recognized as competitively disadvantaged individuals. Thus, it might be the differential interaction of these alternative genotypes with changing ecological and socioeconomic conditions that are producing the observed fertility differentials, as well as the resulting changes in prevalence of psychopathology, and not higher education *per se.*

Although this distinction may seem like hair-splitting, there are important differences in the social policy implications of these different interpretations. Discouraging *K*-selected women from pursuing higher education, increased social status, and enhanced material resources, would be catastrophic for them under current social circumstances if it adversely affected their economic competitiveness without in any way enhancing their fertility, which might be a genetically determined trait that is not very susceptible to social alteration. It would completely negate the expected fitness benefits of their own reproductive strategy of delayed fertility and higher parental investment. In short, careful attention to the "endophenotype" underlying higher education may be critically important in correctly interpreting the observed phenomena.

The fourth chapter in this set, by Underwood, examines both the short-term and long-term fitness consequences of twinning in a Micronesian island population. The lifetime and intergenerational fertility of Chamorro women living in Guam are tracked for nearly half a century. Careful attention is paid to potentially confounding factors, such as relative rates of maternal mortality, length of reproductive careers, offspring sex-ratios, and proportions of offspring surviving to reproductive age. The basic result is that the mothers of twins appear to enjoy a slight selective advantage over the mothers of singletons, even when controlled for all these potentially mitigating factors. It is also instructive to see here how this selective advantage looms so much larger when *not* controlled for these other factors. Although the different reproductive strategies represented here do not have exactly equal fitness payoffs, even in the longer run, the longer-term fitness differentials are nowhere near the immediately apparent reproductive benefits of twinning. This is the kind of demographic analysis that is needed to validly compare the relative performance of alternative reproductive strategies.

The fifth chapter in this set, by MacMurray *et al.*, also considers the reproductive consequences of twinning. Like the previous Comings chapter, it examines the correlation between increased fertility (due in this case to twinning) and a genetically transmitted disorder. The genetic disease targeted is preeclampsia, or pregnancy-induced hypertension (*PIH*), which is affected by the gene encoding endothelial nitric oxide synthase (*NOS3*). This gene is therefore correlated with both higher fertility and a potentially life-threatening biomedical disorder. The chapter reports that prior to the development of a medical treatment for *PIH*, this condition could result in a severely increased rate of both maternal and fetal mortality. Thus, any fitness benefits of carrying this gene were previously offset by a massive fitness cost. This prevented any net selective advantage from accruing to the deleterious allele in the past. The implication is that with modern medical care, we are now inadvertently favoring the replication of the gene for *PIH* by facilitating the accrual of the fitness benefits of the increased twinning without accrual of the fitness costs due to increased mortality. This is precisely the same argument as was used in the assessment of the relative expected costs and benefits of high-fecundity, low parental-investment strategies under varying environmental conditions of resource competition. In this case, using a specific gene with specific biomedical and reproductive consequences, we can more clearly see the fitness tradeoffs involved and the precise conditions that might confer a selective advantage to one or the other alternative. This enhanced degree of precision adds greater support to the fundamental argument made by Comings in the previous chapter, in spite of the relative weakness of the higher education hypothesis, which might otherwise distract one from the main point being made.

Finally, it is interesting to note how these examples illustrate the basically amoral and value-free nature of natural selection. Whereas from the perspective of 19th Century Spenserian Social Darwinism, natural selection was always for the good, and the losers in free market competition were expected to accept their preordained fate with good grace, we now understand that these blind evolutionary processes do not always yield the most socially desirable outcomes. This realization serves as good protection against the "naturalistic fallacy" and should help us to design better social policy than a blind optimistic faith in the redeeming power of natural selection might otherwise produce.

REFERENCES

Gangestad, S.W., & Simpson, J.A. (1990). Toward an evolutionary history of female sociosexual variation. *Journal of Personality, 58,* 69-96.

Hermstein, R.J., & Murray, C. (1994). *The bell curve: Intelligence and class structure in American life.* New York, NY: Free Press.

Rowe, D.C., Vazsonyi, A.T., & Figueredo, A.J. (1997). Mating effort in adolescence: Conditional or alternative strategy? *Journal of Personality and Individual Differences, 23(1),* 105-115.

Rushton, J.P. (1985). Differential K theory: The sociobiology of individual and group differences. *Journal of Personality and Individual Differences, 6,* 441-452.

INDEX